Wearing Gauss's Jersey

Wearing Gauss's Jersey

Dean Hathout

CRC Press
Taylor & Francis Group
Boca Raton London New York

CRC Press is an imprint of the
Taylor & Francis Group, an **informa** business
AN A K PETERS BOOK

MIX
Paper from responsible sources
FSC® C014174
www.fsc.org

CRC Press
Taylor & Francis Group
6000 Broken Sound Parkway NW, Suite 300
Boca Raton, FL 33487-2742

Printed on acid-free paper
Version Date: 20130204

International Standard Book Number-13: 978-1-4665-0864-4 (Hardback)

Library of Congress Cataloging-in-Publication Data

Hathout, Dean.
Wearing Gauss's jersey / Dean Hathout.
pages cm
"An A K Peters book."
Audience: Grade 9 to 12.
Includes bibliographical references.
ISBN 978-1-4665-0864-4 (alk. paper)
1. Gaussian quadrature formulas. 2. Mathematics--Study and teaching (Middle school) 3. Mathematics--Study and teaching (Secondary) I. Title.

QA299.4.G3H38 2013
510--dc23 2012044432

Visit the Taylor & Francis Web site at
http://www.taylorandfrancis.com

and the CRC Press Web site at
http://www.crcpress.com

Contents

List of Problems

Preface

Yesterday, my older brother and I went out to the backyard to play some "one-on-one" basketball. Before the game, another in our heated rivalry, I went into my room and put on my number 24 Kobe Bryant basketball jersey, which I wear for all of our big games. This urge, to pretend to be a sports star or a superhero, is apparently very common among kids. My father, whom I cannot imagine playing make-believe, told me that as a kid he would pretend to be James Bond, Batman, and Daniel Boone (who?), with special outfits, toy guns, and gadgets for each.

A question I recently began pondering is, Why don't kids ever pretend to be famous scientists or mathematicians—why are there no Albert Einstein jerseys or Leonard Euler jerseys? Although it might sound crazy, I think it would be a good idea. So much so, that I had my own jersey made up, with the logo below

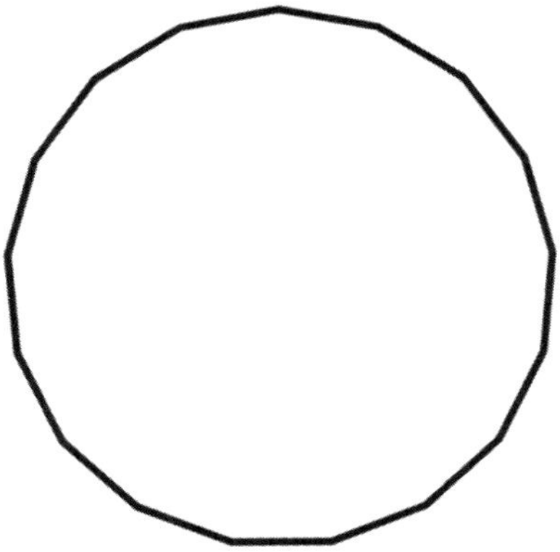

Do you recognize this figure? It is a heptadecagon—a 17-sided regular polygon. This figure is often associated with the name of Carl Friedrich Gauss (1777–1855), one of the world's most brilliant mathematicians. So brilliant, in fact, that he has earned the nickname *Princeps mathematicorum* (Latin for "Prince of Mathematicians"). At the age of 19, Gauss proved that a regular heptadecagon can be constructed by using just a compass and a straightedge (a ruler without numbers). This was the first advance on this problem since the time of the ancient Greeks. For thousands of years, mathematicians have been concerned with the problem of which regular polygons could be constructed by using just a compass and a straightedge (but not, for example, a protractor to measure angles or a ruler to measure side lengths). For example, can a pentagon be made this way? How about a heptagon (a seven-sided figure)?

For a figure to be constructible in this fashion, the compass and straightedge have to be manipulated to construct certain angles without being able to measure them with a protractor. If you have never done this, it is actually quite tedious—I had to do it for geometry class a couple of years ago and truly disliked it.

The idea is that if a central angle of the polygon (or its sine or cosine) can be written out as a number that involves the sums and products of integers and radicals, then it can be constructed. At the age of 19, Gauss proved that a heptadecagon could be so constructed, citing this equation:

$$\cos\left(\frac{2\pi}{17}\right) = -\frac{1}{16} + \frac{1}{16}\sqrt{17} + \frac{1}{16}\sqrt{34 - 2\sqrt{17}} + \frac{1}{8}\sqrt{17 + 3\sqrt{17} - \sqrt{34 - 2\sqrt{17}} - 2\sqrt{34 + 2\sqrt{17}}}.$$

Later, he went on to prove that a regular polygon with n sides can be constructed with compass and straightedge if n is the product of a power of 2 and any number of distinct Fermat primes.

A Fermat prime is a prime number that can be written as

$$F_n = 2^{2^n} + 1.$$

This means that the nth Fermat number is 2 raised to the power 2^n, plus 1. If this number turns out to be prime, it is called a Fermat prime. This looks complicated, but it's actually not so bad. If $n = 0$, then $2^n = 1$, so $2^{2^n} = 2^1$; thus,

$$F_0 = 2^1 + 1 = 3.$$

If $n = 1$, then $2^n = 2$, so $2^{2^n} = 2^2$; thus, $F_1 = 2^2 + 1 = 5$. These are both prime numbers.

Believe it or not, so far only five Fermat primes are known:

$$F_0 = 3,\ F_1 = 5,\ F_2 = 17,\ F_3 = 257, \text{ and } F_4 = 65537.$$

The next 28 Fermat numbers, F_5 through F_{32}, are known to be composite.

Thus, using the rules found by Gauss, an n-gon is constructible if

$$n = 3, 4, 5, 6, 8, 10, 12, 15, 16, 17, 20, 24, \ldots,$$

whereas an n-gon is not constructible with compass and straightedge if

$$n = 7, 9, 11, 13, 14, 18, 19, 21, 22, 23, 25, \ldots.$$

The bottom line is that even though all even multiples of constructible polygons can be constructed (because once you have a square, the angles can be bisected to an octagon and so forth, and it is easy to bisect an angle with a compass and a straightedge), so far, only 31 polygons with odd n can be constructed. Only 31 because only five Fermat primes are known (whether there are more is an open question in number theory), and there are only 31 ways to get distinct products of these Fermat primes.

For now, what we need to know is that this issue is complicated, and, at only 19 years of age, Gauss solved a problem that had puzzled mathematicians for thousands of years before him. He was so pleased with his discovery that the regular heptadecagon could be constructed with a compass and straightedge that he asked that a heptadecagon be carved on his tombstone when he died. Actually, the stone mason refused because he said it would look too much like a circle, so it never ended up on Gauss's tombstone. So, in honor of Gauss, I made my Gauss jersey with the heptadecagon and the formula for Fermat primes.

My interest in Gauss, however, stems from another very famous story about him. When he was a young child, perhaps just 10 years of age, his teacher, a Master Buttner, is reported to have given the class the following problem: calculate the sum of the numbers 1 through 100. Master Buttner apparently intended to keep the children busy for an hour or so doing the lengthy calculation. To his surprise, the young Gauss had the correct answer within just a few seconds. How did he do it?

Gauss had a special insight, which will be the subject of the first problem in the book. More important, though, it is the theme for the book.

This is a book about what we'll call "Gauss problems." In some sense, these are problems just like that given to the young Gauss by Master Buttner—problems that, if tackled in the straightforward, obvious way, can be very tedious and time consuming but, if approached in a more insightful fashion, can yield the solution much more easily and elegantly.

This is what distinguishes this book from the many wonderful problem books already available. Problem books that contain Olympiad problems, or AMC (American Math Competition) problems, obviously have very beautiful mathematics. However they do not focus specifically on "Gauss problems." Nor, in fact, do books on problem-solving techniques, although they come much closer to that spirit.

But why focus on Gauss problems?

Before we go on, let's look at an interesting problem that might help to illustrate and exemplify the notion of mathematical insight, and to differentiate it from problems that are directly obvious, on the one hand, and those that require heavily specialized training (without which no amount of insight can help), on the other.

We start with a very well-known problem among mathematicians. Take an 8×8 grid, such as the one below, and assume that we have a large supply of 2×1 dominoes, each of which can cover exactly two squares on the grid. Is it possible to completely tile the grid with dominoes so that there are no overhangs or overlaps? If so, in how many ways can this be done?

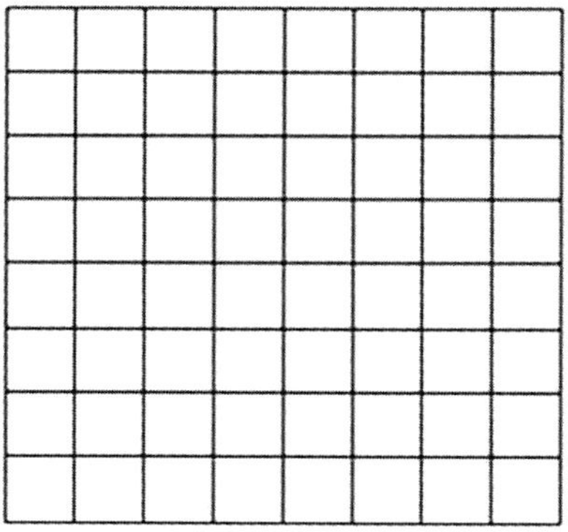

The first question is quite easy. Each domino covers two squares, and we see that there would be many ways to use 32 dominoes to completely cover the grid. This really requires no special insight.

The second question, regarding just how many ways this can be done, turns out to be extremely difficult for us beginners. The answer is 12,988,816 ways. The solution to this problem was published in 1961 by a Dutch physicist named Peiter Willam Kasteleyn and by British mathematicians Michael E. Fisher and Harold N. Temperley, who derived that for a $2m \times 2n$ rectangular grid (using $2mn$ dominoes) there would be

$$4^{mn}\prod_{j=1}^{m}\prod_{k=1}^{n}\left(\cos^2\frac{j\pi}{2m+1}+\cos^2\frac{k\pi}{2n+1}\right).$$

ways to tile the grid.

Therefore, for an 8×8 grid, $m = n = 4$, and there are

$$4^{16}\prod_{j=1}^{m}\prod_{k=1}^{n}\left(\cos^2\frac{j\pi}{9}+\cos^2\frac{k\pi}{9}\right)=12{,}988{,}816$$

different tilings. This problem is a serious question in research mathematics; it requires not just mathematical insight, but a great deal of specialized knowledge and training.

With this introduction, let's imagine that we cut one corner square away from the top of the grid. Is it still possible to tile the grid with dominoes subject to the same rules as before? That is an easy question, since there are now 63 squares in the grid, and each domino covers exactly 2 squares, so there is no way to tile a grid with an odd number of squares.

At last, we arrive at a curious and interesting question. Let us imagine that we cut out the bottom right and top left squares from the grid, so that it now looks like the figure below.

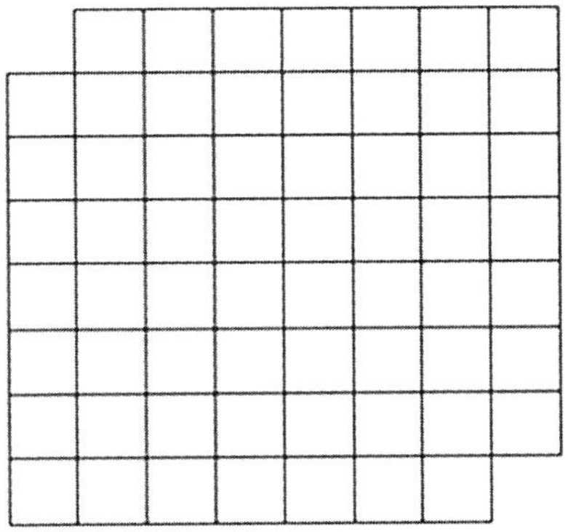

Now there are 62 squares in the grid. How many ways are there to tile this grid with 31 dominoes? As we start to tinker with this problem, trying various arrangements of dominoes, we realize that this might be a rather difficult problem as well. We may even spend many hours trying to find the tiling. If there are almost 13 million tilings for 64 squares, there certainly should be some tilings now, since there are 62 squares, and these should be coverable by 31 dominoes. We could save ourselves those many hours, however, and the accompanying frustration as we waste time unsuccessfully looking for even one tiling, by using a simple, yet brilliant insight. Instead of looking at our original plain grid, let us color it with alternating black and white squares, like a chessboard, as shown below.

When we remove the two corner squares, all of a sudden we realize that we have removed two white squares from opposing corners.

We now have 32 black squares and 30 white squares. When we look at how a 2 × 1 domino covers the board, we see that each domino covers one black square and an adjacent white square.

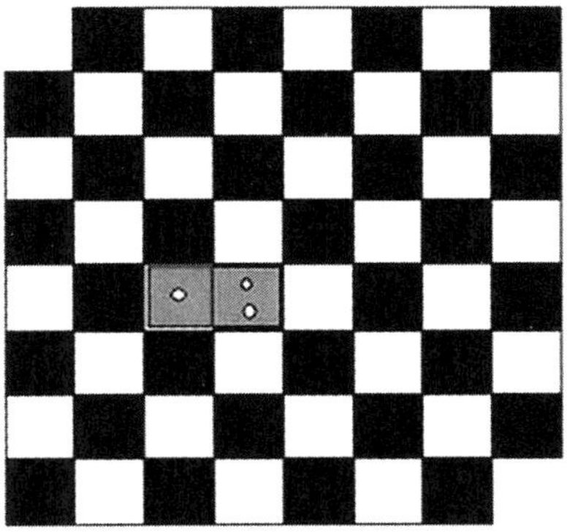

Therefore, whatever number of dominoes we put down, they will cover an equal number of black squares and white squares. Thus, if we are to tile the board with 31 dominoes, we would need 31 black squares and 31 white squares, but we don't have that. Therefore, we have just proven that it is impossible to tile the grid with opposite corners removed. The insight here was very clever, and it involved just a subtle shift of perspective: to see the 8×8 grid as a black and white chessboard! Once the solution is explained to us, it seems both ingenious and elementary at the same time.

That is what this book is about. The exhilarating feeling that comes from seeing that insight, or that very fun feeling of saying, "*How did I miss that*?" I think I have learned most of my math from just that feeling, after failing to solve a problem, or worse yet, solving it in a very cumbersome way, and then reading or seeing a very elegant, insightful solution. The "How did I *not* see that?" feeling is truly exhilarating, and pushes me to try to internalize the insight I just learned so I can use it in another problem.

In my own mind, I think about mathematical insights as being of three different types. The first, which we can call a type 1 insight, is something that requires no special mathematical training or knowledge at all, but relies on pure creativity. This is like the grid problem—a Ph.D. in math might miss it, but a middle-schooler might see the critical insight. Of course, deriving the formula for the full chessboard that we saw above will probably require a Ph.D. in math, and no middle-schooler can come up with that! The second sort of insight, what we can call type 2, is the sort of insight that requires familiarity with some mathematical tool, which creative thinking then helps in cracking a problem. If we don't have that tool at hand (or in mind), then we probably would not be able to have the needed insight (which is to use that particular tool on that particular problem). Therefore, I think studying math really helps in our ability to develop insights. Finally, for type 3 insights, we use trial and error or casework to investigate a problem and then end up recognizing some sort of pattern from our limited cases, which we then go on to prove holds for the entire problem.

Of course, the most fun problems are those that require only type 1 insights, but to be good problem solvers, we'll need to master all three types, and we'll look at examples of each in this book.

The book is divided into chapters, each built around a broad theme, with some "Gauss problems" that illustrate the subject matter of that chapter. These start off easy, just to demonstrate the needed insight to solve the given problem more efficiently. We consider these insights to be our mathematical tools. The beginning problems in each chapter develop simple but powerful tools. Many of the problems end with a short summary of the tools used and new tools developed to solve the problems. We then use these simple tools for two ends: to solve more complex problems along the same theme, and at the same time to develop more complex mathematical tools. Even with fairly simple tools, we see that we can begin to develop our own insights and solve serious problems, such as Olympiad-level problems, which we tackle in the last chapter.

The largest chapters investigate sequences and series (the sort of problem for which the young Gauss is so well known), and combinatorics. These topics are essentially all of the type that would be very tedious to do in a straightforward way but have easier ways of tackling them. The entire field of combinatorics is essentially about finding ways of counting things more efficiently than by counting directly. In fact, one of my favorite books is *Mathematics of Choice* by Ivan Niven (1965). The subtitle of that book is *How to Count without Counting*. Thus, this entire book could have been made up of series problems and combinatorics problems. However, we have tried to show that the same idea of insights applies in algebra, trigonometry, number theory, and a general area called "changing perspectives," which shows how problems in one area (e.g., algebra) can be transformed to a different area (e.g., geometry), making these problems easier to solve.

As the chapters progress, the tools developed in one chapter will be needed to solve problems in seemingly unrelated chapters. That will be another theme that the book seeks to illustrate: there is an underlying unity or harmony in math, and problems that may seem on the surface to be totally unrelated will turn out to be extremely connected to each other.

I hope that, at the end, the reader will arrive at two conclusions:

1. Math is fun!
2. Regardless of whether one's initial level of mathematical insight is inborn or acquired, it can be improved.

Of course, for seasoned mathematicians, most of the problems we cover are elementary and the concepts in combinatorics and trigonometry are rudimentary. We all have to start somewhere, however, and as a young person who likes math, I think it is best to start with the basics—that is how I learn best, and this book is written for people like me.

We keep hearing in the news about how children in the United States are so far behind their counterparts in other nations in the area of mathematics. I think part of the reason may be that math is not presented as a fun subject in school. I am confident, though, that there are many kids out there just like me. We love math, and want to get better at it! We may not now be at the level of Math Olympians (I know I am not), but we can enjoy math just as much and would be proud to do math wearing Gauss's jersey.

1

Arithmetic and Geometric Series

Problem 1.1

Find the sum of the first 100 integers,

$$S = 1 + 2 + 3 + ... + 100.$$

I think that this problem needs to be at the beginning of any book on insightful or "Aha!" mathematical solutions. The problem is not difficult—it is just laborious. We just start adding, and keep a close tally: $1 + 2 = 3$, $3 + 3 = 6$, $6 + 4 = 10$, and so on.

Given half an hour or so, and hoping we don't make a mistake, we should be able to get the answer.

However, what if we have to add up the first 1,000 integers? Or the first 1,000,000?

Insight

As we must have guessed, there is a clever solution to this problem, which, in my opinion, captures the essence of mathematical insight more than any other. Moreover, this solution was supposedly discovered by a 10-year- old, leading to one of the most famous stories in mathematical lore. Here is how the story goes.

A young schoolboy named Carl Friedrich Gauss was given this problem by his teacher, who assigned it to the class in an effort to keep the pupils busy for an hour or so. They each had a piece of chalk and a tablet (small chalkboard) on which to tally the numbers. To the teacher's great chagrin, Gauss walked up in a couple of minutes with the correct answer written on his tablet. How did he do it?

Gauss had the following insight. Let us "duplicate" the problem by writing the same set of numbers both forward and backward:

$$S = 1 + 2 + 3 + 4 + \dots + 100,$$

$$S = 100 + 99 + 98 + 97 + \dots + 1.$$

Since it does not matter (by the commutative property of addition) in what order we add the numbers, we will add the numbers vertically first and then sum horizontally.

$$\begin{array}{rcccccccccc} S & = & 1 & + & 2 & + & 3 & + & 4 & + \dots + & 100 \\ +S & = & 100 & + & 99 & + & 98 & + & 97 & + \dots + & 1 \\ \hline 2S & = & 101 & + & 101 & + & 101 & + & 101 & + \dots + & 101 \end{array}$$

Here, we have added the two expressions for S, summing the first terms $(1 + 100)$, then the second terms $(2 + 99)$, then the third terms $(3 + 98)$, and so on. Each of these pairs of terms adds to 101. If we now add all of these 101s to each other, we have, in effect, added all of the numbers 1, 2, 3, …, 100 to each other *twice*. As we can see, there are 100 pair sums, each of which is 101, and we can say

$$2S = 100 \times 101,$$

or

$$S = \frac{100 \times 101}{2} = 5050.$$

We can easily generalize this problem by using Gauss's method.

$$\begin{array}{rcccccccccc} S & = & 1 & + & 2 & + & 3 & + & 4 & + \dots + & n \\ +S & = & n & + & n-1 & + & n-2 & + & n-3 & + \dots + & 1 \\ \hline 2S & = & n+1 & + & n+1 & + & n+1 & + & n+1 & + \dots + & n+1 \end{array}$$

We can see that there are n groups of $(n + 1)$ because one group corresponds to each of the columns labeled 1 through n. Therefore, we can say

$$2S = n\,(n + 1),$$

and finally, dividing by 2, we have our formula for S:

$$S = \frac{n(n+1)}{2}.$$

In summation notation, then, we have the equation

$$\sum_{i=1}^{n} i = \frac{n(n+1)}{2}.$$

The summation or "sigma" notation means that we do the operation inside sigma as a function of the counting index i, and then add up each value for the specified limits. In this case, the function in terms of i is just i, so when $i = 1$, we list 1, then we add the value of i when $i = 2$, and so forth.

This formula for the sum of the first n natural numbers and the thinking process involved in deriving it will be invaluable in future problem solving.

Going Deeper

Before leaving this problem, let us look at an extension. Try the following problem:

$$S = 3 + 8 + 13 + 18 + \ldots + 108.$$

We now have a good feel for how to tackle this problem. Let us use Gauss's duplication strategy and write the sum twice, forward and backward:

$$S = 3 + 8 + 13 + 18 + \ldots + 108,$$

$$S = 108 + 103 + 98 + 93 + \ldots + 3.$$

We could see that our sequence started with the number 3, and each time, it was incremented by 5 to produce the next term.

If we add the two sequences, we see that each column of two terms adds to 111:

$$2S = 111 + 111 + 111 + 111 + \ldots + 111.$$

Our only issue now is to find how many terms there are.

We know that we can express the last term as $3 + n(5) = 108$. Thus, $n = 21$, and S can be written as

$$S = 3 + 0(5) + 3 + 1(5) + 3 + 2(5) + \ldots + 3 + 21(5).$$

Thus, there are 22 terms ($n = 0$ through and including 21), and we can say

$$2S = 22(111)$$

$$S = \frac{22(111)}{2} = 11(111) = 1{,}221.$$

We can quickly generalize this method to sum any arithmetic progression. This is a sequence of numbers that can be written as

$$S = a + a + d + a + 2d + a + 3d + \ldots + a + (n-1)d.$$

Here, a is called the starting value and d is the common difference between the terms. We can see that the sequence has n terms (0 through $n - 1$).

Let's now use our Gaussian insight to get a formula for the sum of our general arithmetic progression. We first write

$$S = a + a + d + a + 2d + \ldots + a + (n-1)d$$

$$S = a + (n-1)d + a + (n-2)d + a + (n-3)d + \ldots + a$$

Adding vertically and then horizontally, we have

$$2S = 2a + (n-1)d + 2a + (n-1)d + 2a + (n-1)d + \ldots + 2a + (n-1)d.$$

Since there are n terms, we can say

$$2S = 2an + n(n-1)d$$

$$S = \frac{2an + n(n-1)d}{2} = an + \frac{n(n-1)d}{2}.$$

By using this formula, we can now quickly calculate our previous sum,

$$S = 3 + 8 + 13 + 18 + \ldots + 108$$

Here, $a = 3$, $d = 5$, and $n = 22$. Thus,

$$S = 3(22) + \frac{22(21)5}{2} = 66 + 1{,}155 = 1{,}221,$$

just as we had before.

Tools Used and Developed

- Gaussian duplication trick.
- Summation formula for the first n natural numbers:

$$\sum_{i=1}^{n} i = \frac{n(n+1)}{2}.$$

- Summation formula for an arithmetic progression:

$$S = a + a+d + a+2d + a+3d + \ldots + a+(n-1)d,$$

$$S = \frac{2an + n(n-1)d}{2} = an + \frac{n(n-1)d}{2}.$$

Problem 1.1a (extension of Problem 1.1)

To conserve the contents of a 16-ounce bottle of tonic, a castaway adopts the following procedure. On the first day, he drinks 1 ounce of tonic and then refills the bottle with water; on the second day, he drinks 2 ounces of the mixture and then refills the bottle with water; on the third day, he drinks 3 ounces of the mixture and again refills the bottle with water. The procedure is continued for succeeding days until the bottle is empty. How many ounces of water does he thus drink?

Let's give this one some thought.

Insight

We know that the man will drink for 16 days because on day 1, he drinks 1 ounce, on day n he drinks n ounces, and the bottle is a 16-ounce bottle.

After this, it is very easy to get bogged down in this type of problem. The initial temptation is to try to work out a formula for the daily ratios of tonic to water or of water to the mixture, and track how this changes with time. This is very complex, and also very unnecessary. All we have to do is keep track of the water he adds, because we know that eventually, when the bottle is empty, he drank it all. We can think of him refilling the bottle just before he drinks (not just after). Thus, on the morning of day 1, he drinks 1 ounce of tonic. On the morning of day 2, he refills the bottle by adding 1 ounce of water. Then, he drinks 2 ounces of the mixture. On the morning of day 3, he refills the bottle by adding 2 ounces of water. Then he drinks 3 ounces of the mixture, and so on. Thus, on day n, he adds $(n - 1)$ ounces of water.

Since he drinks for 16 days, we know that on the morning of the 16th day, he will have added 15 ounces of water. Then, he will drink all 16 ounces, and the bottle will finally be empty. Thus, keeping track of the ounces of water added, we see that the total amount of water is $1 + 2 + 3 + 4 + \ldots. + 15$. That is now easy for us—we have just done this problem! It is

$$\frac{1}{2}(15)(16) = 120 \text{ ounces.}$$

Tools Used and Developed

- Summation formula for the first n natural numbers,

$$\sum_{i=1}^{n} i = \frac{n(n+1)}{2}.$$

Message

It is unlikely that someone will stop us and say, "Hey kid, can you sum the first n natural numbers?" It is much more likely that this tool will be used as part of some clever puzzle or as part of the solution to an interesting problem.

Problem 1.1b (extension of Problem 1.1)

A group of men working together (i.e., starting together and finishing together) at the same rate can finish a job in 45 hours. However, the men instead begin to work one at a time at equal intervals over a period of time. Once on the job, however, each man stays until the job is finished. If the first man works five times as much as the last man, find the number of hours the first man worked.

How do we approach a problem like this? There is very little information. We don't know how many men there are, nor at what rate they work, nor how long the whole job takes—all seemingly critical pieces of information.

The best approach in these cases is to adopt some good notation, start assigning variables, and write down what we know. Let's give it a go.

Insight

Let us call our variable of interest x, how long the first man works. And let us say there are n men on the job. If the men all work together for the same amount of time, then it takes them 45 hours to do the job. Therefore, the total number of man-hours, call it h, required to complete the job can be expressed as $h = 45n$.

In the second scenario, where the men start at staggered intervals, the first man works for x hours, and the remaining men show up at equal intervals and all stay until the job is done. Say the second man shows up d hours after the first man starts; therefore, he works for $(x - d)$ hours. The third man shows up d hours after the second man arrives, so he works for $(x - 2d)$ hours, and so on. We can see that the last man to arrive, man n, will work for $[x - (n - 1)d]$ hours. If we add up all these hours, they would need to equal $45n$, the total man-hours needed to complete the job.

Let us call S the total hours worked by all the men as they arrive in a staggered fashion. We therefore have

$$S = x + [x - d] + [x - 2d] + [x - 3d] + \ldots + [x - (n - 1)d].$$

We see that this sum has n terms, starting with x and ending with $x - (n - 1)\,d$. Thus, there are n xs, and we could rewrite the sum as

$$S = nx - [d + 2d + 3d + \ldots + (n-1)d] = nx - d\,[1 + 2 + 3 + \ldots + (n-1)].$$

Now we are on some familiar territory. We have the sum of the first $(n-1)$ integers, and we know how to do that very well by now. Thus,

$$S = nx - d\frac{(n-1)n}{2}.$$

But we also know that S equals $45n$, so we have

$$45n = nx - d\frac{(n-1)n}{2},$$

or

$$x - d\frac{(n-1)}{2} = 45. \tag{1.1}$$

We now need our other provided piece of information: the first man works five times as long as the last man. In terms of our variables, this means: $x = 5[x - (n-1)d]$. Let's solve this equation for d in terms of x to get $d = 4x/[5(n-1)]$. If we plug this into Equation (1.1), we get

$$x - \frac{4x}{5(n-1)}\frac{(n-1)}{2} = 45$$

$$\frac{3x}{5} = 45,$$

$$x = 75 \text{ hours}.$$

Amazingly, there is only one possible answer to this problem, which is that the first man works for 75 hours. Interestingly, the remaining variables, such as the total hours worked h, the number of men n, and the separation between their shift starts d, are not uniquely determined. However, if we require these to all be integral, we can get some nice conditions on them from Equation (1.1), now that we know x.

We know that $x - d(n-1)/2 = 45$; plugging in 75 for x, we have $(n-1)d = 60$. Thus, there are many possible combinations for n and d, but they have to meet this condition. For example, $(n - 1) = 4$ and $d = 15$ works. Thus, $n = 5$, and the total number of hours h is $45n = 225$ hours. The five men would then work 75, 60, 45, 30, and 15 hours respectively, which totals 225 hours.

Alternatively, $(n - 1)$ could equal 5 and d could be 12. Thus, there are $n = 6$ men, and the job requires $45(6) = 270$ hours, with the men working 75, 63, 51, 39, 27, 15 hours, respectively, which adds up to 270 hours.

Notice that in both of these scenarios (as well as in all other possibilities) the first man works 75 hours and the last man works 15, fulfilling the requirement that the first man works five times as long as the last man, as given in the problem statement.

It is interesting that even though h, n, and d can vary within bounds, x is uniquely determined. We were able to solve for x, even given the minimum information available.

Tools Used and Developed

- Once again, the key tool here, after correctly setting up equations, was our summation formula,

$$\sum_{i=1}^{n} i = n(n+1)/2.$$

As we said before, no one will say, "Hey kid, can you sum the first n integers?" However, knowing this handy formula will open up the solution to many interesting problems for which it is the key tool.

Problem 1.2

Calculate the following sum:

$$S = 10^2 - 9^2 + 8^2 - 7^2 + 6^2 - 5^2 + 4^2 - 3^2 + 2^2 - 1^2.$$

Bashing It Out

This is easy—almost insulting, we're thinking. So let's just do it:

$$S = 100 - 81 + 64 - 49 + 36 - 25 + 16 - 9 + 4 - 1.$$

Taking the numbers in pairs and doing the subtractions, we get

$$S = 19 + 15 + 11 + 7 + 3 = 55.$$

Now let's do the sum for 100:

$$S = 100^2 - 99^2 + 98^2 - 97^2 + \ldots + 4^2 - 3^2 + 2^2 - 1^2.$$

Uh-oh!

Getting Some Insight

We can see that our prior approach, "just bashing it out," will quickly become impractical. We need a more efficient method. One thing that catches our attention is that the terms can be paired as differences of squares, or

$$S = (100^2 - 99^2) + (98^2 - 97^2) + \ldots + (4^2 - 3^2) + (2^2 - 1^2). \qquad (1.2)$$

We recall that a difference of squares, $a^2 - b^2$, can be rewritten as $(a + b)(a - b)$. This simple factorization can be a great help in many problems. Let us apply this to each of the parentheses in Equation (1.2), and rewrite S as

$$S = (100 + 99)(100 - 99) + (98 + 97)(98 - 97) + \ldots + (4 + 3)(4 - 3) + (2 + 1)(2 - 1).$$

We can see that each right-hand set of parentheses in the pair reduces to 1; for example, (100 – 99) = 1, and so on. Thus, the expression reduces to

$$S = (100 + 99) + (98 + 97) + \ldots. + (4 + 3) + (2 + 1).$$

Now we can remove the parentheses, and we get Gauss's childhood problem,

$$S = 1 + 2 + 3 + \ldots. + 100,$$

which we have previously solved. It was just in disguise! Thus,

$$S = 100^2 - 99^2 + 98^2 - 97^2 + \ldots\ldots + 4^2 - 3^2 + 2^2 - 1^2 = 5{,}050.$$

We can therefore see that one simple maneuver, factoring a difference of squares, reduced a problem that we did not know how to solve efficiently to one with which we are familiar, and we can now solve it easily. That sort of maneuver, to borrow a phrase from Professor Paul Zeitz, can be called the "crux move" in solving a problem. Paul Zeitz, by the way, is a math professor at the University of San Francisco and a former coach of the US International Mathematical Olympiad Team. He has written a wonderful book on problem-solving strategies called *The Art and Craft of Mathematical Problem Solving* (Zeitz, 1999).

We can now generalize this problem to say that for all even n,

$$n^2 - (n-1)^2 + (n-2)^2 - (n-3)^2 + \ \dots \ + 2^2 - 1^2 = \frac{n(n+1)}{2}. \qquad (1.3)$$

But what if n is odd? That's actually not so difficult. We already know how to solve the problem for even n, so we'll start there:

$$S = n^2 - (n-1)^2 + (n-2)^2 - (n-3)^2 + \dots - 2^2 + 1^2.$$

Now the sequence ends with $+ 1^2$ because there are an odd number of terms.

Let us now isolate the terms from $(n - 1)^2$ to 1^2, since there are an even number of terms, and rewrite S as

$$S = n^2 - [(n-1)^2 - (n-2)^2 + (n-3)^2 + \dots + 2^2 - 1^2].$$

The terms in the brackets are once again a problem that we know how to solve, and that it equals $(n-1)(n)/2$, where we have substituted $(n - 1)$ for n in Equation (1.3). So now we can write

$$S = n^2 - \frac{(n-1)(n)}{2} = \frac{2n^2 - (n^2 - n)}{2} = \frac{n^2 + n}{2} = \frac{n(n+1)}{2}.$$

Thus, we get the same formula whether n is even or odd.

Tools Used and Developed

- Difference of squares: $a^2 - b^2 = (a + b)(a - b)$.
- Summation formula for the first n natural numbers,

  $$\sum_{i=1}^{n} i = n(n+1)/2,$$

  which we developed in Problem 1.1.

- Pairing of terms. This can be an extremely powerful tool that we will see again and again. A clever pairing of terms often cracks a difficult problem.
- Even simple tools that we develop in easy or well-known problems can be used to solve more interesting and harder problems. This is a main theme of this book: we attempt to develop tools and insights in some problems and try to put those together to solve harder problems. Beginners like us can solve hard problems—even problems that appear in Mathematical Olympiads.

Problem 1.3

Evaluate the following sum:

$$1 + 3 + 5 + 7 + 9 + \ldots + 19.$$

Brute Force

We see that we are being asked to sum the odd numbers from 1 to 19. We can just evaluate the sum by directly adding up the terms. If we do that, we get $1 + 3 = 4$, $4 + 5 = 9$, $9 + 7 = 16$, and so on. At the end, if we have not made a mistake, we would get the correct answer: 100.

By now, though, we've learned that the brute force approach should be an approach of last resort because for harder problems, such as $1 + 3 + 5 + 7 + 9 + \ldots. + 99$, it quickly becomes impractical. Let's try to use some insight from what we've learned so far.

Insight 1

We can use Gauss's duplication trick:

$$S = 1 + 3 + 5 + 7 + 9 + \ldots + 19,$$

$$S = 19 + 17 + 15 + 13 + 11 + \ldots + 1.$$

Adding these two expressions for S together vertically and then horizontally, we get

$$2S = 20 + 20 + 20 + 20 + 20 + \ldots + 20.$$

We see that each pair of terms sums to 20. How many pairs are there? There are 20 numbers from 1 to 20, half of them even and half of them odd. We are summing just the odd numbers, so there are 10 pairs. So

$$2S = 10(20) = 200$$

$$S = 100.$$

Thus, we get the correct answer. However, if we stop here, we may have missed a critical insight about the nature of the sum of the first n odd integers. This is one of those important formulas or tools that come up again and again in problem solving.

Insight 2

To develop this tool, we need to recognize that we are dealing here with a very simple arithmetic progression of the type we discussed in Problem 1.1:

$$S = a + a + d + a + 2d + a + 3d + \ldots + a + (n-1)d.$$

In this case, a, the starting value, is 1, and d, the common difference, is 2. We already know how to sum an arithmetic progression in general:

$$S = \frac{2an + n(n-1)d}{2} = an + \frac{n(n-1)d}{2}.$$

By plugging in $a = 1$ and $d = 2$, we get

$$S = n + \frac{n(n-1)2}{2} = n + n^2 - n = n^2.$$

Thus, we have arrived at the remarkable formula that the sum of the first n odd numbers is just n^2. In our problem, indeed, when we added the first 10 odd numbers, we got $10^2 = 100$.

Insight 3

Looking back at our brute force solution, we actually had a hint of this beautiful formula. In evaluating the first few partial sums, we said, "We get $1 + 3 = 4$, $4 + 5 = 9$, $9 + 7 = 16$, and so on." Notice that 1 is a perfect square.

Then when we added 1 and 3 (the first two odd numbers), we got 4, or 2^2. Then when we added 5 to that, we got 9, or 3^2. Then when we added the fourth odd number, 7, to the total, we got 16, or 4^2.

Noticing this allows us to develop, or discuss, an example of a critical tool, called mathematical induction. It is a very powerful way to prove things, particularly if we have noticed or guessed what a formula might be, and now we want a way to prove it.

The idea behind mathematical induction is simple: it is like the game where we set up a row of dominoes, and knock the first one down, then it knocks the second, and as the second falls over, it takes down the third, and so on. What we do is develop a mathematical domino effect that will prove a statement out to infinity.

This occurs in two steps:

1. Show that the statement we want to prove is correct for a base case, some number n. Often, n is just 1.
2. Show that if the statement is true for n, then it must be true for the next number, $n + 1$.

These two steps combine to achieve the following: the statement is true for the base case, say, for $n = 1$. Because of statement 2, it must be true for the next case, $n = 2$. Now that we have proven that it is true for $n = 2$, statement 2 automatically proves that it is true for $n = 3$, and so on out to infinity. Thus, we have proven it to be true for all n.

Let's apply this to our problem of summing the first n odd numbers, where we believe, based on our observation of the first few cases, that the sum will be n^2.

1. Let $n = 1$, and we want to show that our conjecture is true. The sum of the first odd number, 1, is just 1, which is 1^2. We can even let the base case be $n = 2$. The sum of the first two odd numbers, $1 + 3 = 4$, is just 2^2. In either case, we have proven that our conjecture is correct for some small base case.
2. Now for the difficult part: to show that if the statement is true for n, then it must be true for $n + 1$. We assume that the statement is true for n; that is,

$$1 + 3 + 5 + \ldots + (2n - 1) = n^2. \tag{1.4}$$

   Here, $(2n - 1)$ is just a shorthand way of writing the nth odd number. For example, if $n = 3$, then $2n - 1 = 5$, which is the third odd number. Once again, we are assuming that this is true for n, and want to prove that it is true for $n + 1$. In other words, if we now add the $(n + 1)$th odd number, the sum of the first $n + 1$ odd numbers will be $(n + 1)^2$.

Since the nth odd number is $(2n - 1)$, the next odd number after that is larger by 2, or $(2n + 1)$. We can also get this by substituting $n + 1$ for n in the formula for the nth odd number (remembering that n is just an index for counting) as follows: $(2(n + 1) - 1) = (2n + 2 - 1) = (2n + 1)$.

Let's now add the $(n + 1)$th odd number, $(2n + 1)$, to both sides of Equation (1.4), or

$$1 + 3 + 5 + \ldots + (2n - 1) + (2n + 1) = n^2 + (2n + 1).$$

Removing the brackets on the right-hand side, we get

$$1 + 3 + 5 + \ldots + (2n - 1) + (2n + 1) = n^2 + 2n + 1.$$

We recognize that $n^2 + 2n + 1$ is just $(n + 1)^2$, and now we can complete the proof by writing

$$1 + 3 + 5 + \ldots + (2n - 1) + (2n + 1) = (n + 1)^2.$$

We went through this proof in excruciating detail, but the idea of mathematical induction is so important that it is worth feeling comfortable with it. We showed that our conjecture was true for $n = 1$ (and $n = 2$). Then, we showed if it is true for some n, it must be true for $n + 1$; therefore, our conjecture that that the sum of the first n odd numbers will be n^2 must be true for $n = 3$. But then, by the same reasoning, if it is true for $n = 3$, it must be true for $n = 4$, and so on out to infinity. This is a great example of proof by induction.

Finally, it is important to realize that there are often many ways to solve a problem. We proved the same fact by using the formula for the sum of an arithmetic progression and also by using the method of mathematical induction.

Tools Used and Developed

- Gaussian pairing tool.
- Formula for the sum of an arithmetic progression,

$$S = \frac{2an + n(n-1)d}{2} = an + \frac{n(n-1)d}{2}.$$

- Formula for the sum of the first n odd integers, $1 + 3 + 5 + \dots + (2n - 1) = n^2$.
- The method of mathematical induction.

Problem 1.4

The four configurations below, labeled as Figures 0, 1, 2, and 3, consist of 1, 5, 13, and 25 non-overlapping unit squares, respectively. If the pattern were continued, how many non-overlapping unit squares would there be in Figure 100?

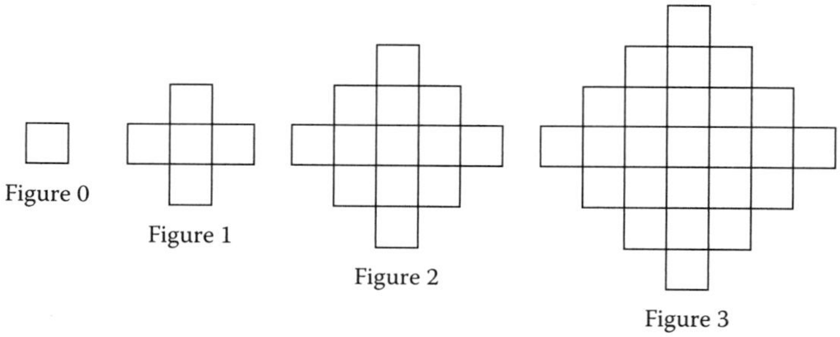

The following solution is the "harder" solution. It is presented here verbatim as it appears in J. Douglas Faires' excellent book, *First Steps for Math Olympians* (Faires, 2006):

> There are numerous ways to recognize a pattern in the figures. The method we have chosen is to first consider the number of additional small squares that would be needed to make the figure a large square. If this completion were done, each large square would consist of $2n + 1$ small squares, for a total of $(2n + 1)^2$ small squares. However, to make the nth figure a square of this size we would need to add small squares to each of the corners. As shown, the number of small squares we would need to add to each of the four corners of the nth figure is
>
> $$1 + 2 + 3 + 4 + \dots + n = \frac{n(n+1)}{2}.$$

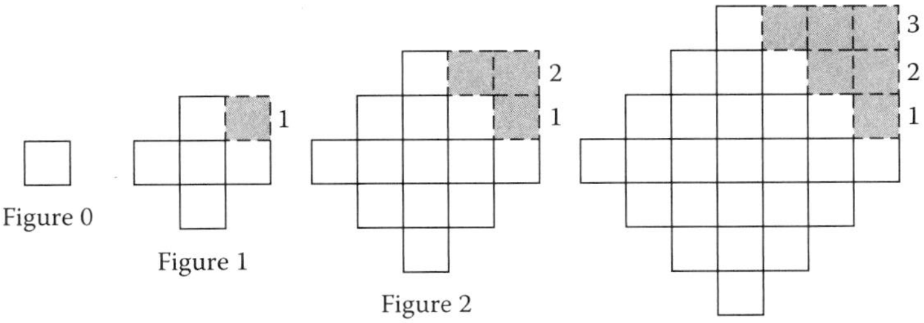

Figure 0 Figure 1 Figure 2 Figure 3

Hence, the number of small squares in the nth figure must be

$$(2n+1)^2 - 4\frac{n(n+1)}{2} = 4n^2 + 4n + 1 - 2n^2 - 2n = 2n^2 + 2n + 1.$$

When $n = 100$, this gives a figure with $2 \times 100^2 + 2 \times 100 + 1 = 20{,}201$ small squares.

Of course, this solution makes nice use of the formula for the sum of the first n natural numbers, which we developed in Problem 1.1. However, it is possible to use another of our tools (and insight) to come up with a solution that is a bit more direct.

Insight

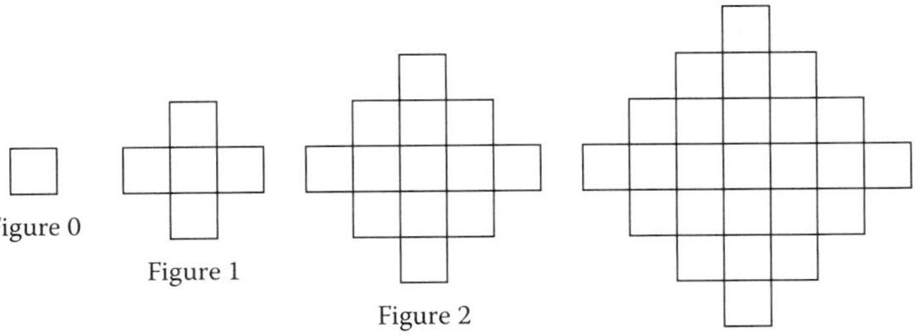

Figure 0 Figure 1 Figure 2 Figure 3

Let us consider each of the four configurations in the original figure to be made up of two adjoining pyramids of squares, one pointing up and one pointing down. For example, in Figure 2, the pyramid pointing up has 3 rows, made up of 1, 3, and 5 squares. The pyramid pointing down has 2 rows, made up of 1 and 3 squares. The top pyramid thus has 1 + 3 + 5 = 9 squares, and the bottom pyramid has 1 + 3 = 4 squares, for a total of 13 squares.

Following this pattern, we can see that each row has an odd number of squares. In Figure n, the up-pointing pyramid will have $n + 1$ rows, and the number of squares in the up-pointing pyramid consists of the sum of the first $n + 1$ odd numbers, whereas the down-pointing pyramid has n rows and is made up of the sum of the first n odd numbers.

Using the results from Problem 1.3, we see that the sum of the first $(n + 1)$ odd numbers is $(n + 1)^2$ and the sum of the first n odd numbers is n^2. Thus, Figure n will have

$$(n + 1)^2 + n^2 = n^2 + 2n + 1 + n^2 = 2n^2 + 2n + 1.$$

As before, when $n = 100$, this gives a figure with $2 \times 100^2 + 2 \times 100 + 1 =$ 20,201 small squares.

This is, therefore, a one-step solution using the formula for the sum of the first n odd numbers.

Tools Used and Developed

- Formula for the sum of the first n odd integers, $1 + 3 + 5 + \ldots .. + (2n - 1) = n^2$, which was developed in Problem 1.3.

Problem 1.5

Sum all of the digits in the numbers 1 through 100. In other words, add up all the digits that make the numbers separately, so that for 11 we add (1 + 1), for 87 we add (8 + 7), and so on, using all the numbers from 1 through 100.

Of course, we know how to do this directly, but it would be very tedious and we don't want to lose track. For 1 through 13, for example, we would add

$$1 + 2 + 3 + 4 + 5 + 6 + 7 + 8 + 9 + (1 + 0) + (1 + 1) + (1 + 2) + (1 + 3) = 55.$$

But we sure don't want to waste the time needed for 1 through 100, so let's try Gauss's trick; we show some of the array below:

1	2	3	…	10	11	12	…	29	30	…	100
100	99	98	…	91	90	89	…	72	71	…	1

Now let's sum the digits of the numbers vertically:

$$1 \text{ and } 100 = 1 + 1 + 0 + 0 = 2,$$

$$2 \text{ and } 99 = 2 + 9 + 9 = 20,$$

$$3 \text{ and } 98 = 3 + 9 + 8 = 20,$$

$$10 \text{ and } 91 = 1 + 0 + 9 + 1 = 11,$$

$$12 \text{ and } 89 = 1 + 2 + 8 + 9 = 20.$$

This is not going well—we see some sort of pattern, but the sums are not consistent, and this does not seem to be as useful an approach as it was when Gauss summed the integers 1–100. We need an insight!

Insight 1

Let us look at the one's digit—we know that it keeps repeating from 0 to 9, regardless of the tens digit. In fact, it will do this ten times, as we break the integers up into groups of ten: (0–9), (10–19), (20–29), etc. Therefore, if we just add up all the ones digits, we have

$$10\sum_{i=0}^{9} i$$

(i.e, 10 times the sum of the digits 0 through 9). Since 0 doesn't contribute to the sum, this can be rewritten as

$$10\sum_{i=1}^{9} i.$$

Now let's look at the tens digits. We will have ten 1s (10–19), ten 2s (20–29), and so on. Therefore, we would have

$$\sum_{i=1}^{9} 10i$$

(i.e., the sum of $10 \times 1 + 10 \times 2 + 10 \times 3 + \ldots$). However, this is the same as

$$10\sum_{i=1}^{9} i,$$

because we can just pull out the 10, or regroup the sum into the sum of (1 + 2 + 3 + … + 9) summed ten times. If we add up both sums, we get

$$10\sum_{i=1}^{9} i + 10\sum_{i=1}^{9} i = 20\sum_{i=1}^{9} i.$$

Since the young Gauss already established that

$$\sum_{i=1}^{n} i = n(n+1)/2,$$

our answer is:

$$20\frac{9(9+1)}{2} = 20\times 45 = 900.$$

Therefore, the sum of all the digits of the numbers 1–100 seems to be 900. Oops! We forgot the number 100 itself! We accounted for all the ones digits and all the tens digits (i.e., all the digits from 0 to 99), but we forgot that there is a single hundreds digit in the number 100. This contributes a 1 to the total sum, and therefore, the correct answer is actually 901.

Now that we have the method, let's try to sum the digits in all of the numbers from 1 through 1,000. We have already done a lot of the hard work. We can divide the numbers into groups of 100: (0–99), (100–199), (200–299), and so on. Here, the sum of the tens and ones digits together, as we have just shown, is 900. We have ten of these groups. Then we have the hundreds digits to think about. There will be one-hundred 1s (100–199), one-hundred 2s (200–299), and so on. Therefore, we would have

$$10\times 900 + \sum_{i=1}^{9} 100i = 9{,}000 + 100\sum_{i=1}^{9} i = 9{,}000 + 100\times 45 = 13{,}500.$$

However, this accounts for only 0 – 999. The number 1,000 contributes a 1 to the sum, so we have 13,501.

Insight 2

As the numbers become larger, our first insight, although useful, becomes more cumbersome. We need a better insight. Let us return to our first instinct, which was to use Gaussian pairing. What hurt us there was that as we shifted from a 9 to a 10, the digit sums did not go in an orderly progression. For example, 99 + 2 and 100 + 1 both sum to 101, but the digit sum in the first case is 20, whereas in the second case, it is 2.

However, this situation can be remedied if we pair the numbers differently, such as the following pairings:

0	1	...	10	11	12	...	29	30	...	45	...
99	98	...	89	88	87	...	70	69	...	54	...

If we pair the numbers as $(a, 99 - a)$, then the ones digits always sum to 9, and the tens digits always sum to 9. Therefore, the digit sum of each pair is always 18. If we look at 0–99, we can divide these numbers into 50 pairs of the form $(a, 99 - a)$. Thus, the sum of the digits from 0 through 99 is $50 \times 18 = 900$. We then add 1 for the digit sum contributed by 100, and we get our original answer of 101 in no time at all. Compare this method, which allowed us to calculate the digit sums of 0–99 as $50 \times 18 = 900$, to our prior method, which resolved, after some slick maneuvering, to $20 \times 45 = 900$.

Our second insight can be easily generalized to sum the digits of any series of the form $1, 2, 3, \ldots, 10^n$. We simply group the numbers into pairs of the form $(a, 10^n - 1 - a)$. In this case, each pair will have a digit sum of $9n$, and there will be $10^n/2$ pairs. Thus, from 0 to $10^n - 1$, the numbers will have a total digit sum of $9n \times 10^n/2$. To get the last number, 10^n, we just add a 1. Therefore, the sum of the digits of all the numbers from 1 to 10^n is $9n \times (10^n/2) + 1$.

Let us try this for our previous calculation for 1–1,000. Here, $n = 3$, so we would get $9(3) \times (10^3/2) + 1 = 27 \times 500 + 1 = 13{,}501$. This is a *much* faster and slicker way, and it used our Gaussian pairing tool. We knew Gauss wouldn't let us down!

Tools Used and Developed

- Gaussian pairing.
- The idea that we may need to pair numbers creatively to solve a problem.

Problem 1.6

Evaluate the following sum:

$$1+\frac{1}{2}+\frac{1}{4}+\frac{1}{8}+\frac{1}{16}+\ldots+\frac{1}{256}.$$

This is a classic problem that may look familiar, and it leads to some rather important ideas.

Brute Force

This problem is not too difficult, and can be done with brute force (which we are learning by now should only be a last resort). We can get a common denominator of 256, add the terms up, and get 511/256. To sum more terms, the calculation would get time consuming quite quickly, and we need an insight. Moreover, if we are asked to sum the series above for an infinite number of terms 1 + 1/2 + 1/4 + 1/8 + 1/16 + …, then brute force is entirely useless.

Insight

Let's generalize the problem a bit by recognizing that each term in the series is of the form $1/2^n$. Let us call the sum we want S, and write the series as:

$$S=1+\frac{1}{2}+\frac{1}{4}+\frac{1}{8}+\frac{1}{16}+\ldots+\frac{1}{2^n}. \tag{1.5}$$

This is called a geometric series, which is defined as a series in which the ratio of succeeding terms is constant. In other words, we can get the next term from the previous one, in this case by multiplying by the common ratio 1/2.

This gives us the idea that, since all the terms are related by this common ratio, we should multiply S in Equation (1.5) by 1/2:

$$\frac{1}{2}S=\frac{1}{2}+\frac{1}{4}+\frac{1}{8}+\frac{1}{16}+\frac{1}{32}+\ldots+\frac{1}{2^{n+1}}. \tag{1.6}$$

Now let's subtract Equation (1.6) from Equation (1.5), and then add, writing the result as

$$S-\frac{1}{2}S=\left(1-\frac{1}{2}\right)+\left(\frac{1}{2}-\frac{1}{4}\right)+\left(\frac{1}{4}-\frac{1}{8}\right)+\left(\frac{1}{8}-\frac{1}{16}\right)$$
$$+\left(\frac{1}{16}-\frac{1}{32}\right)+\ldots+\left(\frac{1}{2^n}-\frac{1}{2^{n+1}}\right). \tag{1.7}$$

We see a wonderful thing happening: the –1/2 in the first parentheses will cancel with the 1/2 in the second parentheses; the –1/4 in the second parenthesis will cancel with the 1/4 in the third parentheses; the –1/8 in the third parenthesis will cancel with the 1/8 in the fourth parentheses; and so on. Thus, the sum will collapse in on itself. Mathematicians call this "telescoping"—a term coined from those old spyglass telescopes that would collapse on themselves when pushed in. The only terms left in Equation (1.7) will be 1 and $-1/2^{n+1}$. We then have

$$\frac{1}{2}S=1-\frac{1}{2^{n+1}}$$

$$S=2-\frac{2}{2^{n+1}}=2-\frac{1}{2^n}.$$

We can use this formula for our original problem,

$$S = 1+\frac{1}{2}+\frac{1}{4}+\frac{1}{8}+\frac{1}{16}+\ldots+\frac{1}{256}.$$

We see that the last term, 1/256, can be written as $1/2^8$. Thus, our sum is

$$S=2-\frac{1}{2^8}=\frac{511}{256}.$$

We can also answer the question of what happens when there are an infinite number of terms in this geometric series. As n tends to infinity, $1/2^n$ tends to 0, so our sum is

$$S=1+\frac{1}{2}+\frac{1}{4}+\frac{1}{8}+\frac{1}{16}+\ldots=2.$$

That's a nice result: the sum of an infinite series of fractions, all reciprocals of the powers of 2, actually adds up to 2.

Now let us completely generalize the problem to sum a general geometric series by using what we've learned:

$$S = a + ar + ar^2 + ar^3 + ar^4 + ... + ar^n.$$

In this geometric series, we see that the common ratio is r. So, we do what we did before, and multiply S by r, then subtract.

$$S = a + ar + ar^2 + ar^3 + ar^4 + ... + ar^n$$

$$rS = ar + ar^2 + ar^3 + ar^4 + ar^5 .... + ar^{n+1}.$$

If we subtract the bottom equation from the top, we get the same cancellations we did before, and the result is

$$S - rS = a - ar^{n+1}$$

$$(1-r)S = a(1-r^{n+1})$$

$$S = \frac{a(1-r^{n+1})}{(1-r)}.$$

This is the general formula for the sum of a geometric series.

Now we can think about an infinite geometric series. If the common ratio r has an absolute value *less than 1*, then as n goes to infinity, r^{n+1} goes to 0, and the series will converge to

$$S = \frac{a}{(1-r)}.$$

If, however, the common ratio r has an absolute value *greater than or equal to 1*, then the infinite series obviously diverges.

Tools Used and Developed

- Pairing of terms, developed in Problem 1.1.
- Formula for the sum of a geometric series

$$S = a + ar + ar^2 + ar^3 + ar^4 + ... + ar^n = \frac{a(1-r^{n+1})}{(1-r)}.$$

- Formula for the sum of a convergent infinite geometric series, $S = a/(1 - r)$.
- The telescoping tool.

Problem 1.7

Does the infinite series $1 + 2(\frac{1}{2}) + 3(\frac{1}{2})^2 + 4(\frac{1}{2})^3 + 5(\frac{1}{2})^4 + \ldots$ converge? If so, find the limit to which it converges.

This is a bit of a tough one if we haven't seen something like it before. We know from Problem 1.6 that the ordinary geometric series $1 + (\frac{1}{2}) + (\frac{1}{2})^2 + (\frac{1}{2})^3 + (\frac{1}{2})^4 + \ldots$, converges. However, after the first term, this new series is bigger, on a term-by-term basis, than the ordinary geometric series with a common factor of $\frac{1}{2}$. Therefore, does it still converge, or not?

One way to go is to start taking partial sums of the series

$$\sum_{i=1}^{n} n(\tfrac{1}{2})^{n-1}$$

for larger and larger values of n, and see if it seems to grow quickly or not. This is very painful to do, and also it may not be conclusive. For example, the harmonic series, $1 + \frac{1}{2} + \frac{1}{3} + \frac{1}{4} + \frac{1}{5} + \ldots$, is well known to diverge, but grows extremely slowly, so that taking partial sums may give the false impression that it is actually a convergent series. So we need an insight.

Insight

Let us tackle the problem in general. We assume that $1 + x + x^2 + x^3 + x^4 \ldots$ is a convergent geometric series. Then, we know that its sum is $1/(1 - x)$. Let's look at the series

$$S = 1 + 2x + 3x^2 + 4x^3 + 5x^4 + \ldots.$$

We already know a good trick in dealing with geometric series, so let's use it. We'll multiply both sides by x to get

$$S = 1 + 2x + 3x^2 + 4x^3 + 5x^4 + \ldots$$

$$xS = \quad x + 2x^2 + 3x^3 + 4x^4 + \ldots.$$

Now, when we subtract the second equation from the first, we get

$$(1-x)S = 1 + x + x^2 + x^3 + x^4 \ldots.$$

Wow! How about that? The sum on the right is just our convergent geometric series. Thus, we get

$$S = \frac{1}{(1-x)^2}.$$

Thus, when $x = 1/2$, as in this problem, $S = 1/[(1 - 1/2)^2]$, or $S = 4$. So the series converges, but its sum is bigger than the ordinary geometric series.

Tools Used and Developed

- The sum of a geometric series, $1 + x + x^2 + x^3 + x^4 \ldots$, developed in Problem 1.6.
- The sum of a modified geometric series of the type $1 + 2x + 3x^2 + 4x^3 + 5x^4 \ldots$.

Problem 1.7a (extension of Problem 1.7)

Two idiots, let's call them A and B, decide to have an old-fashioned pistol duel. They have only one gun (a six-shot revolver) and only one bullet. Thus, they agree to duel as follows. They put the one bullet into the gun, and they face off, directly in front of each other. Idiot A begins by spinning the cylinder, and shooting at B (who is directly in front of him, so is impossible to miss). If the gun fires, then A wins. However, since there is only one bullet, chances are the gun doesn't fire. In that case, B takes the gun, and he spins the cylinder, and then attempts to fire at A. They go back and forth like this until one of them is shot and the other is the winner. What is the probability that A wins?

In this problem, the solution, to some extent, depends on finding a good way to look at the problem and to label variables. So we look at all the ways A can win, calculate the probabilities, and then sum these probabilities.

Let us make the following definitions for the critical events from A's perspective: a is the probability that the gun fires for A, and x is the probability that the gun does not fire. Thus, the duels that A can win would go like this:

$$a$$

$$x\,x\,a$$

$$x\,x\,x\,x\,a$$

and so on

Thus, if the gun fires on the first shot (a), then A wins. If it doesn't fire on the first shot, then the only way A can win is if it also doesn't fire during B's turns but then fires on one of A's turns. Notice that A takes the 1st, 3rd, and 5th shots, and so on. That is, if the duel keeps going, he fires only on the odd turns. That explains the configuration of the winning sequences for A above.

Because there is only one bullet in the six shooter, and because A and B always begin by spinning the cylinder to randomize the location of that bullet, we know that the probability P of event a (the gun fires for A) is 1/6. The probability of x (that the gun does not fire on any turn) is 5/6. Thus, the probability that A wins the duel (call this P(A)) is the sum of the infinite series that represents the probabilities of each winning duel for A occurring, or

$$P(\mathrm{A}) = \frac{1}{6} + \left(\frac{5}{6}\right)^2 \frac{1}{6} + \left(\frac{5}{6}\right)^4 \frac{1}{6} + \ldots$$

Thus,

$$P(\mathrm{A}) = \frac{1}{6}\left[1 + \left(\frac{5}{6}\right)^2 + \left(\frac{5}{6}\right)^4 + \ldots\right].$$

The term in brackets is a convergent infinite geometric series (call it S), where $a = 1$, and r, the common ratio, is $(5/6)^2$. From Problem 1.6, we know that $S = a/(1 - r)$. Thus, in this case,

$$P(\mathrm{A}) = \frac{1}{6}\left[\frac{1}{1 - \left(\frac{5}{6}\right)^2}\right] = \frac{1}{6}\left[\frac{1}{1 - \frac{25}{36}}\right] = \frac{1}{6}\left[\frac{36}{36 - 25}\right] = \frac{6}{11}.$$

Thus, the probability that A wins the duel is 6/11, or 0.545454..., which is just over 50%. A has an advantage because he starts first, but personally, I thought that his probability of winning would be much higher before actually calculating it out.

This problem is a very neat application of the tools we have been studying.

Going Deeper

Now that we are comfortable with geometric series, we can go a bit deeper and ask the question: On average, how long would this duel last? In other words, if there were millions of pairs of A and B who each had a duel, how many shot attempts, on average, must be made before the duel ends? As we saw, this sort of duel could, by chance, go on for a very long time. In fact, we can calculate a probability of the duel lasting any number (say, n) of shot attempts. It is simply $(5/6)^{n-1}(1/6)$ because the gun would not fire for the first $n-1$ shot attempts, and then fire on the nth shot attempt, ending the duel.

In mathematical language, the question we are asking is: If D is the duration of the duel (measured in shot attempts), what is the *expected value* of D? The "expected value" of a variable x, $E(x)$, is just another way of expressing the average value of x. It is calculated by multiplying each possible value of x by its probability and adding these products. In terms of equations,

$$E(x) = \sum_{i}^{n} x_i P(x_i),$$

where x_i represents each of the possible values of x, and $P(x_i)$ is the probability that that value occurs.

Now, for our problem, the probability that the duration of the duel is k shot attempts is given by $P(D = k) = (5/6)^{k-1}(1/6)$, as we have already discussed. Thus,

$$E(D) = \sum_{k=1}^{\infty} kP(D = k)$$

$$= \sum_{k=1}^{\infty} k\left(\frac{5}{6}\right)^{k-1}\frac{1}{6} = \frac{1}{6}\left[1 + 2\left(\frac{5}{6}\right) + 3\left(\frac{5}{6}\right)^2 + 4\left(\frac{5}{6}\right)^3 + \ldots\right].$$

We recognize that the sum in the brackets is a modified geometric series of the form

$$S = 1 + 2x + 3x^2 + 4x^3 + 5x^4 + \ldots.$$

Here, $x = 5/6$. As we derived in Problem 1.7, we know that $S = 1/(1-x)^2$, so

$$E(D) = \frac{1}{6}\left(\frac{1}{\left(1-\frac{5}{6}\right)^2}\right) = 6.$$

Thus, the duel, on average, will last for six shot attempts.

Once again, we see that puzzles can be a problem we've already solved, just in disguise!

Tools Used and Developed

- How to sum a geometric series, $1 + x + x^2 + x^3 + x^4 \ldots$, developed in Problem 1.6.
- How to sum a modified geometric series of the type $1 + 2x + 3x^2 + 4x^3 + 5x^4 \ldots.$, developed in Problem 1.7.
- The notion of expected value.

Problem 1.8

Find the following sum:

$$S = \frac{1}{1\cdot 2} + \frac{1}{2\cdot 3} + \frac{1}{3\cdot 4} + \frac{1}{4\cdot 5}.$$

We know how to sum such a series. First, we simplify the denominators in each fraction:

$$S = \frac{1}{2} + \frac{1}{6} + \frac{1}{12} + \frac{1}{20},$$

then we find a common denominator. By inspection, we can see that 60 is the least common denominator, so we rewrite the fractions as

$$S = \frac{30}{60} + \frac{10}{60} + \frac{5}{60} + \frac{3}{60} = \frac{48}{60} = \frac{4}{5}.$$

That wasn't too bad. Now, however, suppose we want to sum

$$S = \frac{1}{1\cdot 2} + \frac{1}{2\cdot 3} + \frac{1}{3\cdot 4} + \ldots + \frac{1}{99\cdot 100}.$$

That's just crazy. It'll take us a couple of hours to simplify all the denominators, find a least common denominator, add everything up, and then simplify the fraction—and we have to do it all without making any mistakes. We need an insight.

Insight 1

Our first insight is of the type where we see a possible pattern—just a glimmer of hope that perhaps there is a simple solution—and it gives us something to investigate. We notice that the last term we added is $1/(4 \times 5)$ and our final answer was 4/5. Coincidence? At this stage, we don't know, but it's worth investigating further. Let's add one more term and evaluate

$$\frac{1}{1\times 2} + \frac{1}{2\times 3} + \frac{1}{3\times 4} + \frac{1}{4\times 5} + \frac{1}{5\times 6}.$$

We already know the sum of the first four terms, so this is really

$$\frac{4}{5} + \frac{1}{5\cdot 6} = \frac{4}{5} + \frac{1}{30} = \frac{24}{30} + \frac{1}{30} = \frac{25}{30} = \frac{5}{6}.$$

We now conjecture that if

$$S = \frac{1}{1\times 2} + \frac{1}{2\times 3} + \frac{1}{3\times 4} + \ldots + \frac{1}{n\times (n+1)},$$

the final answer is $S = n/(n + 1)$. Granted, this conjecture is made on very flimsy grounds, but it gives us something to work from. How would we prove our conjecture? As we've seen before with formulas we believe are valid for all n, if we know that they are true from some base case, we can use mathematical induction. We have already proven our formula for some base cases, so now we want to show that if it is true for n, it will be true for $n + 1$. In our specific situation, we want to show that if

$$\frac{1}{1\times 2}+\frac{1}{2\times 3}+\frac{1}{3\times 4}+\ldots+\frac{1}{n(n+1)}=\frac{n}{n+1},$$

then

$$\frac{1}{1\times 2}+\frac{1}{2\times 3}+\frac{1}{3\times 4}+\ldots+\frac{1}{n(n+1)}+\frac{1}{(n+1)(n+2)}=\frac{(n+1)}{(n+2)}.$$

By the induction hypothesis, we know that the first n terms (through $1/[n(n + 1)]$) sum to $n/(n + 1)$. Therefore, we need to show only that

$$\frac{n}{(n+1)}+\frac{1}{(n+1)(n+2)}=\frac{(n+1)}{(n+2)}.$$

We start by getting a common denominator for the left-hand side,

$$\frac{n(n+2)}{(n+1)(n+2)}+\frac{1}{(n+1)(n+2)}=\frac{n^2+2n+1}{(n+1)(n+2)},$$

which can be written as

$$\frac{n^2+2n+1}{(n+1)(n+2)}=\frac{(n+1)^2}{(n+1)(n+2)}.$$

Canceling $(n + 1)$ from the top and bottom, we get $(n + 1)/(n + 2)$. This completes our proof, and we can now easily answer our harder question, finding

$$S=\frac{1}{1\times 2}+\frac{1}{2\times 3}+\frac{1}{3\times 4}+\ldots+\frac{1}{99\times 100}.$$

We can confidently say that the answer has to be 99/100, without doing hours of tedious work.

Insight 2

Here, we got lucky. Although our solution is perfectly valid, we're still left a bit unsatisfied, as sometimes happens with mathematical induction proofs. We first had to guess the form of the solution, and then we proved it by mathematical induction. What if we hadn't seen that the solution seems to be of the form $n/(n+1)$? Our second insight discusses a more direct assault on the problem.

With series problems, one approach worth investigating is whether we can get the series to *telescope*. What does this mean? It is essentially the pairing tool that we saw in Problem 1.6, where most of the terms in the series cancel each other out, making the solution very simple.

Let's be a little more explicit about what we mean. Let's say we are summing terms in a series, where each term is some function of k, say $g(k)$. We are seeking the sum

$$S=\sum_{k=1}^{n} g(k)=g(1)+g(2)+g(3)+\cdots+g(n).$$

If we are somehow able to rewrite $g(k)$ in such a way that $g(k)=f(k)-f(k+1)$, then we could rewrite S as

$$\sum_{k=1}^{n} g(k)=g(1)+g(2)+\cdots+g(n)$$

$$=[f(1)-f(2)]+[f(2)-f(3)]+[f(3)-f(4)]+\cdots+[f(n)-f(n+1)],$$

where $g(1)=[f(1)-f(2)]$, $g(2)=[f(2)-f(3)]$, and $g(n)=[f(n)-f(n+1)]$.

We can now see what will happen: in the first bracket, we have a $-f(2)$, which cancels the $+f(2)$ in the second bracket, $-f(3)$ in the second bracket cancels the $+f(3)$ in the third bracket, and so on down the line. All of the inside terms cancel (i.e., the series telescopes), and we are left with

$$S=f(1)-f(n+1).$$

The question is: Can we do this in our problem? That is, can we rewrite $g(k)$ in such a way that $g(k)=f(k)-f(k+1)$? Here we have a good shot at it because each $g(k)$ is a fraction with a denominator made up of a multiplication of different factors. In this case, we can try a technique called partial fractions. In this approach, we assume that our fraction can be broken up into two fractions with simpler denominators:

$$\frac{1}{k(k+1)} = \frac{A}{k} + \frac{B}{k+1}.$$

We do not know the values of A and B here, but we are trying to see whether we can determine them. If we can, then we can break up our fraction as above.

For the common denominator, we get

$$\frac{A(k+1)}{k(k+1)} + \frac{Bk}{k(k+1)} = \frac{(A+B)k+A}{k(k+1)}.$$

For this to equal $1/[k(k+1)]$, we need $A = 1$ and $B = -1$. Thus, we have succeeded in expressing each term $g(k) = 1/[k(k+1)]$ as $[f(k) - f(k+1)] = 1/k - 1/(k+1)$, precisely what we needed.

We can check this for a couple of cases, and we see that

$$\frac{1}{2\times 3} = \frac{1}{2} - \frac{1}{3} = \frac{1}{6} \quad \text{and} \quad \frac{1}{3\times 4} = \frac{1}{3} - \frac{1}{4} = \frac{1}{12},$$

as expected. Thus, we can rewrite our sum

$$\frac{1}{1\times 2} + \frac{1}{2\times 3} + \frac{1}{3\times 4} + \ldots + \frac{1}{99\times 100}$$

as

$$\left(\frac{1}{1} - \frac{1}{2}\right) + \left(\frac{1}{2} - \frac{1}{3}\right) + \left(\frac{1}{3} - \frac{1}{4}\right) + \cdots + \left(\frac{1}{99} - \frac{1}{100}\right).$$

We then see the magic of telescoping. All of the inside terms cancel, and we are left with $1/1 - 1/100 = 99/100$, just as we had before.

We can see, of course, that this solution is quite general, and that for the series

$$\frac{1}{1\times 2} + \frac{1}{2\times 3} + \frac{1}{3\times 4} + \ldots + \frac{1}{(n-1)n},$$

the sum will be

$$S = \frac{1}{1} - \frac{1}{n} = \frac{n-1}{n}.$$

Tools Used and Developed

- The idea of recognizing a possible solution pattern from case work.
- Mathematical induction.
- Partial fraction decomposition, such as

$$\frac{1}{k(k+1)} = \frac{A}{k} + \frac{B}{k+1}.$$

- Telescoping a series (i.e., expressing $g(k)$ as some $[f(k) - f(k+1)]$) and transforming the sum as:

$$\sum_{k=1}^{n} g(k) = g(1) + g(2) + \cdots + g(n)$$

$$= [f(1) - f(2)] + [f(2) - f(3)] + [f(3) - f(4)] + \cdots + [f(n) - f(n+1)].$$

Problem 1.9

Find the sum the following series:

$$S = 1(1!) + 2(2!) + 3(3!) + 4(4!) + 5(5!).$$

We know factorial notation, $n! = n(n-1)(n-2)\ldots1$, so $2! = 2$; $3! = 3 \times 2 \times 1 = 6$; $4! = 4 \times 3 \times 2 \times 1 = 24$; $5! = 5 \times 4 \times 3 \times 2 \times 1 = 120$, and we can proceed directly. Notice that we can save ourselves considerable time, since

$$(n+1)! = (n+1)[n(n-1)(n-2)\ldots1] = (n+1)n!$$

Thus, we can calculate 5! simply by doing $5(4!) = 5(24) = 120$, as we saw above. This saves us a bit of time, and after doing the math, we get $S = 719$.

Noticing that we did this fairly quickly, let's now consider the sum $S = 1(1!) + 2(2!) + 3(3!) + \ldots + 10(10!)$. This will start to get painful. We need some sort of insight.

Insight

One great insight is to realize that there is probably nothing special about 10(10!) as a last term. Therefore, one good insight is to generalize the problem:

$$S = 1(1!) + 2(2!) + 3(3!) + \ldots + n(n!).$$

The clever trick we learned for summing a geometric series, multiplying by the common ratio r, and then subtracting S from rS, does not look like it can help us here, though, because there is no common ratio. However, we have seen that another clever general tool that can help us sum a series is telescoping, which makes the terms cancel each other out so the series just collapses. Is it possible to get this series to telescope?

We don't know, but we can begin playing around. Let us explore the time-saving tool we just talked about, that $(n + 1)! = (n + 1)n!$ We write our original series S as the difference of two series:

$$\begin{array}{rcccccccl} & 2(1!) & + 3(2!) & + 4(3!) & + \ldots & + n(n-1)! & + (n+1)n! & & (1.8) \\ - & 1! & + 2! & + 3! & + \ldots & + (n-1)! & + n! & & (1.9) \\ \hline S = & 1(1!) & + 2(2!) & + 3(3!) & + \ldots & + (n-1)(n-1)! & + n(n!) & & \end{array}$$

We see that if we subtract Equation (1.9) from Equation (1.8) term-by-term, we get our series S. For example, if we look at the 3! term, when we do $4(3!) - 3!$, we get $3(3!)$, as in S.

Initially, this may seem like it made matters worse. Now, instead of dealing with one troublesome series, we have to deal with two. However, remember that we did this move for a reason, wanting to exploit the simple identity $(n + 1)! = (n + 1)n!$

Now, looking at Equation (1.8), we can write $2(1!)$ as $2!$, $3(2!)$ as $3!$, and so on, down to $n(n - 1)! = n!$ and $(n + 1)n! = (n + 1)!$ for the last two terms. So we can now rewrite our subtraction as

$$\begin{array}{rcccccc} & 2! & + 3! & + 4! & + \ldots & + n! & + (n+1)! \\ - & 1! & + 2! & + 3! & + \ldots & + (n-1)! & + n! \\ \hline \end{array}$$

Now we can see the brilliance of this approach. We have achieved a telescoping sum. The 2! above will cancel with the 2! below, the 3! above with the 3! below, and so on. All we have left is $(n+1)!$ above and 1! below, and we now have our answer in general:

$$S = 1(1!) + 2(2!) + 3(3!) + \ldots + (n-1)(n-1)! + n(n!) = (n+1)! - 1$$

Let's check it with our first problem: $S = 1(1!) + 2(2!) + 3(3!) + 4(4!) + 5(5!)$. We expect this to equal $6! - 1 = 719$, which is the answer we got the hard way.

For our second problem, summing through $10(10!)$, we know the answer will be $11! - 1$. This is a bit more difficult to calculate, but when we do it, we get $39{,}916{,}800 - 1 = 39{,}916{,}799$.

Now that we have solved our problem, let's reflect a little bit. Certainly, our insight made the problem much easier. However, there was still one hard part—we had to calculate 11! by brute force, and it's a really big number. This makes us think: is there a slick way to calculate factorials, since they get big quite fast? Unfortunately, there is no simple formula to just plug into. However, to pique our interest for later study in mathematics, there is a famous formula that gives a good approximation for factorials of large numbers, known as Stirling's formula:

$$n! \approx \sqrt{2\pi n}\left(\frac{n}{e}\right)^n.$$

The derivation of Stirling's formula is rather difficult, but it can be something to look forward to as we advance in math. The formula does not give an exact answer, but the relative error is less than 1% when $n = 10$, and gets progressively less as n grows.

Tools Used and Developed

- $(n+1)! = (n+1)n!$.
- The telescoping tool.

Problem 1.10

Find the sum of the series

$$S = \frac{1}{2!} + \frac{2}{3!} + \frac{3}{4!} + \frac{4}{5!}.$$

Again, definitely doable. First, let's simplify the denominators in each fraction by calculating the factorials:

$$S = \frac{1}{2} + \frac{2}{6} + \frac{3}{24} + \frac{4}{120}.$$

We need to find a common denominator. We can see that 120 is the least common denominator because it contains all of the other denominators within it—all $k!$, where $k < 5$, so we have

$$S = \frac{60}{120} + \frac{40}{120} + \frac{15}{120} + \frac{4}{120}.$$

Adding these up, we get 119/120. Very interesting!

Now let's look at a bit more complicated problem:

$$S = \frac{1}{2!} + \frac{2}{3!} + \frac{3}{4!} + \cdots + \frac{9}{10!}.$$

Looking at our answer to the first problem, we ask, could we have another simple pattern on our hands? We could proceed as we did for Problem 1.9 by latching onto this pattern and trying to prove it by induction, so that we don't have to do all of the tedious arithmetic. However, is there an even better insight?

Insight

Each term seems to be of the form $k/[(k + 1)!]$. In Problem 1.8, we were able to produce telescoping by using the method of partial fractions. Can we do that again?

One approach is to use the trick that $(k + 1)! = (k + 1)k!$, and break up the fraction as

$$\frac{k}{(k+1)!} = \frac{A}{(k+1)} + \frac{B}{k!}.$$

To get a common denominator of $(k + 1)!$, we have

$$\frac{Ak!}{(k+1)k!} + \frac{B(k+1)}{(k+1)k!}.$$

For this to equal $k/[(k+1)!]$, we need $Ak! + B(k+1) = k$. That doesn't look like it's going to work.

We realize that this attempt was totally misguided anyway. Our task with telescoping is to be able to take some function $g(k)$ and rewrite it in such a way that $g(k) = f(k) - f(k+1)$. Here, $g(k) = k/[(k+1)!]$. However, the two parts that we were trying to decompose it into do not have the same form. One is $f(k) = A/(k+1)$, and the other is a different function entirely, some $h(k) = B/(k!)$. This doesn't help us with telescoping at all.

With this new focus, we might get a flash of inspiration: let's express k as $[(k+1) - 1]$. But is this useless? Even worse, is it stupid? Let's see. We start with

$$\frac{k}{(k+1)!} = \frac{(k+1)-1}{(k+1)!}.$$

Now let's break up the fraction:

$$\frac{(k+1)-1}{(k+1)!} = \frac{(k+1)}{(k+1)!} - \frac{1}{(k+1)!}.$$

All of a sudden, we see it. Since

$$\frac{(k+1)}{(k+1)!} = \frac{(k+1)}{(k+1)\,k!} = \frac{1}{k!},$$

we can write

$$\frac{k}{(k+1)!} = \frac{1}{k!} - \frac{1}{(k+1)!}.$$

Success! We now have our $f(k)$, which is $1/(k!)$, and we have succeeded in expressing $g(k) = k/[(k+1)!]$ as $[f(k) - f(k+1)]$. As before, the telescoping now takes care of itself, because

$$S = g(1) + g(2) + \ldots + g(n) = [f(1) - f(2)] + [f(2) - f(3)] +$$

$$[f(3) - f(4)] + \ldots + [f(n) - f(n+1)].$$

The series will telescope, and we will be left with $S = f(1) - f(n+1)$.

In our problem,

$$S = \frac{1}{2!} + \frac{2}{3!} + \frac{3}{4!} + \cdots + \frac{9}{10!},$$

we know that this will be

$$\frac{1}{1!} - \frac{1}{10!} = 1 - \frac{1}{10!}.$$

For the general case,

$$S = \sum_{k=1}^{n} \frac{k}{(k+1)!} = \frac{1}{2!} + \frac{2}{3!} + \frac{3}{4!} + \cdots + \frac{n}{(n+1)!},$$

we can see that the answer is

$$S = 1 - \frac{1}{(n+1)!}.$$

Tools Used and Developed

- The telescoping tool.
- Being on the lookout for clever ways to reduce $g(k)$ to $[f(k) - f(k+1)]$. In our case, all it took was the small manipulation of rewriting k as $[(k+1) - 1]$.

Problem 1.11

Define a sequence of numbers as follows:

$$a_1 = 2; \quad a_{n+1} = 3a_n + 1.$$

Calculate the sum of the first five terms in this sequence.

This is certainly easy enough. In general, these recursively defined sequences are tedious because we have to calculate one term, then plug it into the definition to get the next, and so on. However, we can just get rolling from the definition provided:

$$a_2 = 3a_1 + 1 = 3(2) + 1 = 7,$$

$$a_3 = 3a_2 + 1 = 3(7) + 1 = 22,$$

$$a_4 = 3a_3 + 1 = 3(22) + 1 = 67,$$

$$a_5 = 3a_4 + 1 = 3(67) + 1 = 202.$$

Therefore, the needed sum is $2 + 7 + 22 + 67 + 202 = 300$.

That wasn't hard at all. Now, though, let's do the same problem for the first 20 terms. This will become mind-numbingly tedious, and we take a risk of making some small mistake along the way, not recognizing it, and carrying it forward, so that all of our effort would be wasted. Therefore, we need an insight.

Insight 1

We are hoping to get a closed-form solution—a formula that will give us the sum of the first *n* terms. We see that the sequence has a fairly simple recursive definition, so there will probably be a pattern as we expand the terms. Our big problem, of course, is that the sequence is not a simple geometric sequence because the ratio of the terms is not constant. But it seems close enough that we can try to find a pattern that we can use.

$$a_2 = 3a_1 + 1$$

$$a_3 = 3a_2 + 1 = 3(3a_1 + 1) + 1 = 3^2 a_1 + 3^1 + 1$$

$$a_4 = 3a_3 + 1 = 3(3^2 a_1 + 3^1 + 1) + 1 = 3^3 a_1 + 3^2 + 3^1 + 1$$

$$\vdots$$

We can now see a pattern, and get the idea to list the terms vertically and get the sum that way.

$$a_1$$

$$3\,a_1 + 1$$

$$3^2 a_1 + 3^1 + 1$$

$$3^3 a_1 + 3^2 + 3^1 + 1$$

$$3^4 a_1 + 3^3 + 3^2 + 3^1 + 1$$

$$\vdots$$

$$3^{n-1} a_1 + 3^{n-2} + 3^{n-3} + \cdots + 1$$

We see that we get the following series of sums:

$$a_1 \sum_{i=0}^{n-1} 3^i + \sum_{i=0}^{n-2} 3^i + \sum_{i=0}^{n-3} 3^i + \sum_{i=0}^{n-4} 3^i + \cdots + \sum_{i=0}^{0} 3^i. \tag{1.10}$$

The last summation in this equation represents the 1 at the end of the last term.

Each of the terms in Equation (1.10) is a geometric series with 3 as the common ratio. We know that such a series,

$$\sum_{i=0}^{m} 3^i,$$

will have a sum of the form $(3^{m+1} - 1)/2$. If we now factor out 1/2 and apply the sum of the series formula to Equation (1.10), we get

$$\frac{1}{2}\left[a_1(3^n - 1) + (3^{n-1} - 1) + (3^{n-2} - 1) + \cdots + (3^1 - 1)\right].$$

Now we can collect terms (especially the −1s), and we get

$$\frac{1}{2}\left[a_1(3^n - 1) + (3^{n-1} + 3^{n-2} + \cdots + 3^1) - (n-1)\right] =$$

$$\frac{1}{2}\left[a_1(3^n - 1) + (3^{n-1} + 3^{n-2} + \cdots + 3^1 + 1) - n\right].$$

Once again, by using the sum of the geometric series in the last set of parentheses, we get

$$\frac{1}{2}\left[a_1(3^n-1)+\frac{1}{2}(3^n-1)-n\right].$$

This now gives us a closed-form solution for the sum we seek. All we need to do is substitute $a_1 = 2$, and we get

$$\frac{1}{2}\left[2(3^n-1)+\frac{1}{2}(3^n-1)-n\right].$$

So we are able to use our knowledge of summing geometric series to get the answer. I must admit, I made about three mistakes in keeping track of all the different terms, and discovered them by using the formula and finding that it gave me the wrong answer when compared to the long-hand calculation of all the terms above.

Insight 2

All of the work we did in the first insight was necessary because the recursive definition was not precisely a geometric series. In other words, there wasn't a constant ratio between successive terms. It sure was close, though. How about if we try to turn the problem into a straight geometric series? How can we do that? Let's look at the following. Let us define $b_n = a_n + 1/2$ for all n, which is the same as

$$a_n = b_n - \frac{1}{2}.$$

Rewriting our recursive definition $a_{n+1} = 3a_n + 1$ in terms of b_n, we get

$$b_{n+1} - \frac{1}{2} = 3\left(b_n - \frac{1}{2}\right) + 1,$$

which simplifies to

$$b_{n+1} = 3b_n.$$

Amazingly, we have reduced the problem to a simple geometric series, which starts with a single term b_1, and then for $n > 1$ is defined by $b_{n+1} = 3b_n$. If we sum the terms of this series, we get $b_1[(3^n - 1)/2]$. From the definition $b_n = a_n + 1/2$, we see that $b_1 = 5/2$, so we get $(5/4)(3^n - 1)$ as our final formula for the sum of the b terms. We have to remember, however, that $b_n = a_n + 1/2$ for all n. Since we summed all the b terms, we added an extra 1/2 with each term over what we would have gotten if we had summed the a terms. After summing n terms, we would thus have added an extra $n/2$, which we now need to subtract. Therefore, the answer to our problem is

$$\frac{5}{4}(3^n - 1) - \frac{n}{2}.$$

If we plug in $n = 5$, we get 300, as we saw before. To get the answer for $n = 20$, we have$(5/4)(3^{20} - 1) - 10 = 4{,}358{,}480{,}490$.

Problem 1.12

Simplify the following product:

$$\left(10^{2^0}+1\right)\left(10^{2^1}+1\right)\left(10^{2^2}+1\right)\left(10^{2^3}+1\right).$$

This is a great way to keep busy because we can waste a lot of time doing arithmetic. But we can simplify directly, beginning by attacking the powers of 10 raised to exponents, and we can rewrite the sum as

$$\left(10^1+1\right)\left(10^2+1\right)\left(10^4+1\right)\left(10^8+1\right).$$

This can now be simplified to 11(101)(10001)(100000001), which gives us 1,111,111,111,111,111. I had to count a few times on the calculator display to make sure I got the number of 1s correct. I think I did.

So now let's calculate

$$\left(10^{2^0}+1\right)\left(10^{2^1}+1\right)\left(10^{2^2}+1\right)\left(10^{2^3}+1\right)\left(10^{2^4}+1\right)\left(10^{2^5}+1\right).$$

This would be really tedious to do directly. Our smaller problem certainly suggests a pattern: the answer will be a string of 1s. However, can we be sure the pattern won't break down? Also, how many 1s will there be if the pattern holds? We need an insight to help us.

Insight

Once again, a great insight is that there is probably nothing special about the last term, and we can try the problem in general for a last term of $(10^{2^n} + 1)$. As we have seen before, one good strategy to try when faced with adding or multiplying a long series of numbers is to try to make the series collapse or somehow telescope. How do we do that? We need a domino effect to make the series collapse. Also, dealing with exponents that are themselves raised to powers can be difficult, so let's play with that a little bit.

Just to get our feet wet, let's look at one term, such as 10^{2^3}, and see what happens when we square it:

$$\left(10^{2^3}\right)^2 = 10^{2^3} \cdot 10^{2^3} = 10^{(2^3+2^3)} = 10^{2(2^3)} = 10^{2^1(2^3)} = 10^{2^{(1+3)}} = 10^{2^4}.$$

Thus, we have the slightly surprising result that $(10^{2^3})^2 = 10^{2^4}$. We might have thought that it would equal 10^{2^6}, but as we can see from our careful work above, it doesn't. Exponents of exponents work in strange ways that may defy our intuition. However, the above exercise was useful in that we learned that 10^{2^k} can produce $10^{2^{k+1}}$ if it is squared. Thus, we can seemingly generate the hard part of one term in our problem from the previous term just by squaring it!

This gives us the idea that since all of the terms are of the form $(a + 1)$, we can find a way to produce a^2 in there somewhere. The easiest way to do this is to multiply $(a + 1)$ by $(a - 1)$ to produce $(a^2 - 1)$.

Getting back to our product of the series of binomials,

$$P = \left(10^{2^0} + 1\right)\left(10^{2^1} + 1\right)\left(10^{2^2} + 1\right)\ldots\left(10^{2^n} + 1\right).$$

We cannot change the value of this product, but we can multiply it by 1 in a creative way that helps utilize our previous results. Since our first factor in P is $(10^{2^0} + 1)$, let us multiply P by

$$\left(\frac{10^{2^0} - 1}{10^{2^0} - 1}\right) = \left(\frac{10^{2^0} - 1}{9}\right)$$

and pull the 1/9 out front, so we have

$$P = \frac{1}{9}\left(10^{2^0} - 1\right)\left(10^{2^0} + 1\right)\left(10^{2^1} + 1\right)\left(10^{2^2} + 1\right)\ldots\left(10^{2^n} + 1\right).$$

Let us focus for a moment on the product of the first two binomials, $(10^{2^0} - 1)(10^{2^0} + 1)$. This is a product of the form $(a - 1)(a + 1) = (a^2 - 1)$, and the result is $(10^{2^0})^2 - 1$. However, we already know that $(10^{2^k})^2 = 10^{2^{k+1}}$, so we see that $(10^{2^0} - 1)(10^{2^0} + 1) = (10^{2^1} - 1)$. That's amazing! Multiplying by the next binomial, $(10^{2^1} + 1)$, the same logic gives us that $(10^{2^1} - 1)(10^{2^1} + 1) = (10^{2^2} - 1)$. This, in turn, will multiply and collapse the next term, $(10^{2^2} + 1)$, yielding the intermediate result of $(10^{2^3} - 1)$. Each intermediate result acts like a domino, collapsing the next term, until we reach the end with the product $(10^{2^n} - 1)(10^{2^n} + 1)$, which will yield $(10^{2^{n+1}} - 1)$. Remembering the 1/9 out front, we finally have that

$$P = \frac{1}{9}\left(10^{2^{n+1}} - 1\right).$$

For our first problem, $(10^{2^0} + 1)(10^{2^1} + 1)(10^{2^2} + 1)(10^{2^3} + 1)$, $n = 3$, so we would have $(1/9)(10^{2^4} - 1)$. In other words, we have 1,111,111,111,111,111, which we got by direct multiplication. For our harder problem, the one with six factors, we would get

$$P = \frac{1}{9}\left(10^{2^6} - 1\right) = \frac{1}{9}\left(10^{64} - 1\right).$$

We know that 10^{64} is 1 followed by sixty-four 0s, so $(10^{64} - 1)$ is the number made by a string of 9s: 999…999 (sixty-four 9s), and when this is multiplied by 1/9, we get our final answer, $P = 111\ldots111$ (sixty-four 1s).

Thus, we see how a clever insight saved us a lot of work. Thus, we can see how very simple tools can produce very nice results—all that is needed is the right insight.

We realize that there is nothing special in this problem about the base 10, and any number could be substituted in its place. The same line of thinking would yield that $(a^{2^0} + 1)(a^{2^1} + 1)(a^{2^2} + 1)(a^{2^3} + 1)$ simplifies to

$$\frac{1}{a-1}\left(a^{2^{n+1}} - 1\right).$$

Tools Used and Developed

- $(a-1)(a+1)=(a^2-1)$.
- A clever way of multiplying by 1, in this case

$$\left(\frac{10^{2^0}-1}{10^{2^0}-1}\right)=1.$$

- The multiplicative telescoping tool.
- $(a^{2^k})^2 = a^{2^{k+1}}$.

Problem 1.13

Find the sum

$$\sqrt{1+\frac{1}{2^2}+\frac{1}{3^2}}+\sqrt{1+\frac{1}{3^2}+\frac{1}{4^2}}+\sqrt{1+\frac{1}{4^2}+\frac{1}{5^2}}.$$

This sequence problem looks more complicated than it is. It is a bit cumbersome, as there is a lot of calculation here, but it's certainly not difficult. When we start to simplify, we get

$$\sqrt{1+\frac{1}{4}+\frac{1}{9}}+\sqrt{1+\frac{1}{9}+\frac{1}{16}}+\sqrt{1+\frac{1}{16}+\frac{1}{25}}$$

$$=\sqrt{\frac{36}{36}+\frac{9}{36}+\frac{4}{36}}+\sqrt{\frac{144}{144}+\frac{16}{144}+\frac{9}{144}}+\sqrt{\frac{400}{400}+\frac{25}{400}+\frac{16}{400}}$$

$$=\sqrt{\frac{49}{36}}+\sqrt{\frac{169}{144}}+\sqrt{\frac{441}{400}}=\frac{7}{6}+\frac{13}{12}+\frac{21}{20}=\frac{198}{60}=\frac{33}{10}.$$

If we keep going, we find the following sum:

$$\sqrt{1+\frac{1}{2^2}+\frac{1}{3^2}}+\sqrt{1+\frac{1}{3^2}+\frac{1}{4^2}}+\sqrt{1+\frac{1}{4^2}+\frac{1}{5^2}}+\cdots+\sqrt{1+\frac{1}{19^2}+\frac{1}{20^2}}.$$

Clearly, this is way too time-consuming, and we need an insight. We think to ourselves, What would Gauss do?

Insight

We noticed in our brute force calculations that at least the common denominator for each of the fractions inside each radical came out to a perfect square, which we would expect since it is the product of two perfect squares (e.g., in the first radical, our common denominator was 36, the product of 2^2 and 3^2. Somehow, though, the numerator also came out as a perfect square, which we really didn't expect. Thus, we are hopeful that there are some insights beneath the surface, and all we have to do is look hard enough to find them.

We start out by looking at the form of each term in the sequence:

$$\sqrt{1+\frac{1}{n^2}+\frac{1}{(n+1)^2}}.$$

When we play with this a bit, we see that we can get a common denominator for the summands under the radical, and that the sum can be expressed as

$$\frac{n^2(n+1)^2+(n+1)^2+n^2}{n^2(n+1)^2}.$$

We try to rewrite the numerator and express the fraction as

$$\frac{n^2[(n+1)^2+1]+(n+1)^2}{n^2(n+1)^2},$$

and then rewrite it as

$$\frac{(n+1)^2[n^2+1]+n^2}{n^2(n+1)^2}.$$

Although we see a certain symmetry in both of these formulations for the numerator, they do not seem particularly helpful, and neither seems to be a perfect square, as our earlier direct case work had suggested.

Therefore, we decide to expand the numerator entirely to see whether that would help.

$$\frac{n^2(n^2+2n+1)+(n^2+2n+1)+n^2}{n^2(n+1)^2}=\frac{n^4+2n^3+3n^2+2n+1}{n^2(n+1)^2}.$$

This last expression gives us an idea. Could $n^4 + 2n^3 + 3n^2 + 2n + 1$ be written as $(n^2 + n + 1)^2$? It certainly seems to make sense, since there would be only one way to get n^4 and only one way to get 1. Also, there would be two ways to get n^3, which would consist of an n^2 from one bracket and an n from the other. Same for n, which we can get in two ways by using an n from one bracket and a 1 from the other. However, there would be three ways to make n^2, which would be an n^2 from one bracket and a 1 from the other (two ways), as well as an n from each bracket. We excitedly check the multiplication and find out that, indeed, we can rewrite our fraction as

$$\frac{n^4+2n^3+3n^2+2n+1}{n^2(n+1)^2}=\frac{(n^2+n+1)^2}{n^2(n+1)^2}.$$

This explains why both the numerator and denominator turned out to be perfect squares when we calculated the terms in the original sequence. Thus, each term in our original sequence can be written as

$$\sqrt{1+\frac{1}{n^2}+\frac{1}{(n+1)^2}}=\sqrt{\frac{(n^2+n+1)^2}{n^2(n+1)^2}}=\frac{(n^2+n+1)}{n(n+1)}.$$

This is amazing progress, which makes us feel that we must be on the right track. We do another simplification:

$$\frac{n^2+n+1}{n(n+1)}=\frac{n(n+1)+1}{n(n+1)}=1+\frac{1}{n(n+1)}.$$

The fraction $1/[n(n + 1)]$ looks wonderfully familiar. We saw it back in Problem 1.8, where we decomposed it into

$$\frac{1}{n(n+1)}=\frac{1}{n}-\frac{1}{n+1}.$$

This is outstanding because, as in Problem 1.8, this sets us up to be able to collapse the series by telescoping. Each term can now be written as

$$\sqrt{1+\frac{1}{n^2}+\frac{1}{(n+1)^2}}=1+\frac{1}{n}-\frac{1}{n+1}.$$

Therefore,

$$\sqrt{1+\frac{1}{2^2}+\frac{1}{3^2}}+\sqrt{1+\frac{1}{3^2}+\frac{1}{4^2}}+\sqrt{1+\frac{1}{4^2}+\frac{1}{5^2}}=$$

$$\left(1+\frac{1}{2}-\frac{1}{3}\right)+\left(1+\frac{1}{3}-\frac{1}{4}\right)+\left(1+\frac{1}{4}-\frac{1}{5}\right).$$

We can see further that this telescopes down to $(1 + 1 + 1 + 1/2 - 1/5) = 33/10$. Most important, we see the pattern: we get a 1 from each term, and then the remaining fractions will telescope. Therefore, for a general sequence such as

$$\sqrt{1+\frac{1}{2^2}+\frac{1}{3^2}}+\sqrt{1+\frac{1}{3^2}+\frac{1}{4^2}}+\sqrt{1+\frac{1}{4^2}+\frac{1}{5^2}}+\cdots+\sqrt{1+\frac{1}{m^2}+\frac{1}{(m+1)^2}},$$

we see that we have $(m - 1)$ terms, so our final sum would be

$$(m-1)+\frac{1}{2}-\frac{1}{m+1}.$$

Therefore, for the sum

$$\sqrt{1+\frac{1}{2^2}+\frac{1}{3^2}}+\sqrt{1+\frac{1}{3^2}+\frac{1}{4^2}}+\sqrt{1+\frac{1}{4^2}+\frac{1}{5^2}}+\cdots+\sqrt{1+\frac{1}{19^2}+\frac{1}{20^2}},$$

we know our answer will be $18 + 1/2 - 1/20 = 369/20$. This was a great problem, as it let us use a lot of our tools.

Tools Used and Developed

- Partial fraction decomposition,

$$\frac{1}{n(n+1)}=\frac{1}{n}-\frac{1}{n+1}.$$

- Series telescoping tool.
- The factorization $n^2(n + 1)^2 + (n + 1)^2 + n^2 = n^4 + 2n^3 + 3n^2 + 2n + 1 = (n^2 + n + 1)^2$, which is what really helped us crack this problem.

Problem 1.14

What is the following sum?

$$S = 1^2 + 2^2 + 3^2 + 4^2 + 5^2.$$

One of the most famous mathematical insights is that of Gauss, when his teacher asked him to sum the numbers 1–100, and he solved this problem in a matter of seconds, recognizing that the formula to sum a series of numbers is $n \times (n+1)/2$. This problem goes to the next level—to sum a series of squares.

With a little bit of grunt work, we can find this by actually computing the square of each number and then summing them together.

$$1^2 + 2^2 + 3^2 + 4^2 + 5^2 = 1 + 4 + 9 + 16 + 25 = 55.$$

Because we are working with just a few small numbers, this was easy enough. However, what if a problem asked us to sum the squares from 1^2 to 100^2? What would we do then? Clearly, we need a Gaussian insight.

Insight

In general, how should we best go about solving

$$\sum_{k=1}^{n} k^2?$$

Our insight, again, comes in the form of a telescoping series. Rather than summing only k^2, we are actually going to sum $(k+1)^3 - k^3$. We choose this sum specifically, recognizing that it will telescope.

Let's take a look at the summation

$$\sum_{k=1}^{n} (k+1)^3 - k^3,$$

which leads to the following expression:

$$\sum_{k=1}^{n} (k+1)^3 - k^3 = \left(2^3 - 1^3\right) + \left(3^3 - 2^3\right) + \left(4^3 - 3^3\right) + \ldots \left[(n+1)^3 - n^3\right].$$

As we can see, this expression telescopes because all of the terms, except for $(n + 1)^3 - 1^3$, end up canceling out. Hence,

$$\sum_{k=1}^{n}(k+1)^3 - k^3 = (n+1)^3 - 1^3.$$

Now that we have found our telescoping series, the next step in this process is to expand $(k + 1)^3 - k^3$. We know that $(k + 1)^3 - k^3 = 3k^2 + 3k + 1$, so we can rewrite our summation as

$$\sum_{k=1}^{n}3k^2 + 3k + 1 = (n+1)^3 - 1^3.$$

Now we can begin to simplify our summation by splitting it up as

$$\sum_{k=1}^{n}3k^2 + \sum_{k=1}^{n}3k + \sum_{k=1}^{n}1 = (n+1)^3 - 1^3.$$

We can simplify this further into

$$3\sum_{k=1}^{n}k^2 + 3\sum_{k=1}^{n}k + \sum_{k=1}^{n}1 = (n+1)^3 - 1^3.$$

Now let's start substituting for the things that we know.

$$3\sum_{k=1}^{n}k^2 + 3\times\frac{(n)(n+1)}{2} + n = (n+1)^3 - 1^3.$$

We are now in a position to combine all of the like terms to figure out what

$$\sum_{k=1}^{n}k^2$$

equals. Let's simplify our summation step-by-step:

$$3\sum_{k=1}^{n} k^2 + 3\times\frac{(n)(n+1)}{2} + n = (n+1)^3 - 1^3$$

$$3\sum_{k=1}^{n} k^2 + 3\times\frac{(n)(n+1)}{2} + n = n^3 + 3n^2 + 3n$$

$$3\sum_{k=1}^{n} k^2 + 3\times\frac{(n)(n+1)}{2} = n^3 + 3n^2 + 2n$$

$$6\sum_{k=1}^{n} k^2 + 3(n)(n+1) = 2n^3 + 6n^2 + 4n$$

$$6\sum_{k=1}^{n} k^2 + 3n^2 + 3n = 2n^3 + 6n^2 + 4n$$

$$6\sum_{k=1}^{n} k^2 + 3n = 2n^3 + 3n^2 + 4n$$

$$6\sum_{k=1}^{n} k^2 = 2n^3 + 3n^2 + n$$

$$\sum_{k=1}^{n} k^2 = \frac{2n^3 + 3n^2 + n}{6} = \frac{n(n+1)(2n+1)}{6}.$$

We thus have our summation equation for the values of k^2, making it very easy to sum a series of squares!

Now let's use the telescoping process to try to derive a formula for summing a series of cubes,

$$\sum_{k=1}^{n} k^3.$$

What telescoping series should we use? Analogous to the case for the sum of squares, let's use $(k + 1)^4 - k^4$. This telescoping series will get rid of all the k^4 terms, and all terms will cancel except $(n + 1)^4 - 1^4$. Let's make sure we see why this is true by considering

$$\sum_{k=1}^{n}\left[(k+1)^4-k^4\right]=\left(2^4-1^4\right)+\left(3^4-2^4\right)+\left(4^4-3^4\right)+\ldots+\left[(n+1)^4-n^4\right].$$

After the telescoping, we have,

$$\sum_{k=1}^{n}\left[(k+1)^4-k^4\right]=(n+1)^4-1^4.$$

We can simplify $(k+1)^4-k^4$ as $4k^3+6k^2+4k+1$, and $(n+1)^4-1^4$ as $n^4+4n^3+6n^2+4n$, making our equation

$$\sum_{k=1}^{n}\left(4k^3+6k^2+4k+1\right)=n^4+4n^3+6n^2+4n.$$

Just as before, we can begin to simplify our summation step-by-step in order to solve for

$$\sum_{k=1}^{n}k^3:$$

$$4\sum_{k=1}^{n}k^3+6\sum_{k=1}^{n}k^2+4\sum_{k=1}^{n}k+\sum_{k=1}^{n}1=n^4+4n^3+6n^2+4n$$

$$4\sum_{k=1}^{n}k^3+6\times\frac{(n)(n+1)(2n+1)}{6}+4\times\frac{(n)(n+1)}{2}+n=n^4+4n^3+6n^2+4n$$

$$4\sum_{k=1}^{n}k^3+(n)(n+1)(2n+1)+2(n)(n+1)+n=n^4+4n^3+6n^2+4n$$

$$4\sum_{k=1}^{n}k^3+2n^3+3n^2+n+2n^2+2n+n=n^4+4n^3+6n^2+4n$$

$$4\sum_{k=1}^{n}k^3+2n^3+5n^2+4n=n^4+4n^3+6n^2+4n$$

$$4\sum_{k=1}^{n} k^3 = n^4 + 2n^3 + n^2$$

$$\sum_{k=1}^{n} k^3 = \frac{n^4 + 2n^3 + n^2}{4}$$

$$= \left(\frac{(n)(n+1)}{2} \right)^2.$$

So we have our formula for summing a series of cubes. With an understanding of telescoping series, we can find a formula to sum the powers of any order.

Tools Used and Developed

- Simplifying summations in order to derive new formulas.
- Forming telescoping series to derive formulas for the summation of squares and cubes.

2

Counting and Combinatorics

Problem 2.1

Alice, Bob, Cindy, and Doug run a 100-meter dash. Assuming no ties, how many different results are possible for this race?

This type of problem is quite easy to understand. To do it directly, we just list the possibilities for first, second, third, and fourth place in the race and then count them up. It takes a little bit of care to avoid getting mixed up, to avoid duplications, and to avoid leaving out any possibilities.

Let's designate the racers by their first initials as {A,B,C,D}, with the following possible orders:

{A,B,C,D},{A,B,D,C},{A,C,B,D},{A,C,D,B},{A,D,B,C},{A,D,C,B},
{B,A,C,D},{B,A,D,C},{B,C,A,D},{B,C,D,A},{B,D,A,C},{B,D,C,A},
{C,A,B,D},{C,A,D,B},{C,B,A,D},{C,B,D,A},{C,D,A,B},{C,D,B,A},
{D,A,B,C},{D,A,C,B},{D,B,A,C},{D,B,C,A},{D,C,A,B},{D,C,B,A}.

As we can see, there are 24 possibilities. These possibilities are called permutations, an arrangement of a set of objects in a particular order. Counting permutations means counting how many different ways objects can be arranged when order matters.

Suppose Evelyn, Frank, George, and Henry join Alice, Bob, Cindy, and Doug for a second race, so we have a total of eight runners. How many possible race results are there now, again assuming no ties? Using our previous notation, and designating the racers by their first initials, what we are really asking for is the number of permutations of {A,B,C,D,E,F,G,H}. Of course, we could list them out and count them, but we can see that this would be both very tedious and very error prone. We need an insight, and this insight is the first step in our study of combinatorics.

Insight

To count all of the permutations of the eight racers, let us think this way: For first place, any of the eight runners could win, so we have eight choices. Once someone has come in first, that runner cannot also come in second, so for second place, we have seven possible choices. The same logic applies for third place because we already have runners in first and second place (say, Cindy first and George second), so there are only six choices for third place. The same logic applies down the line. When we get to last place, after we already have seven runners in first through seventh place, in any given permutation, there is only one choice for last place. Therefore, the total number of permutations is given by:

$$8 \times 7 \times 6 \times 5 \times 4 \times 3 \times 2 \times 1 = 40{,}320.$$

We recognize this product as 8! (called "eight factorial"), which is how we calculate the number of permutations of eight objects. It certainly is very lucky that we didn't try to list and count the possibilities. In general, for n objects when the order of the objects matters, there are $n!$ permutations.

Therefore, we can go back to our original problem of four racers and say that the number of permutations is $4! = 4 \times 3 \times 2 \times 1 = 24$, as we obtained by brute force.

Let's ask a related question. If we don't care about the order of all of the eight racers, but just about the medals for first, second, and third place, then how many possible permutations are there? In other words, what if each race is now defined by the order of those who win the first-, second-, and third-place medals, but we don't care how the last five racers place. How do we calculate these permutations? We use the same logic as before. For first place, there are eight possibilities; then, for second place, there are seven possibilities; and for third place, that would leave six possibilities. Therefore, the total number of permutations now would be: $8 \times 7 \times 6 = 336$. There actually is a nice way of writing this that will help us a lot in combinatorics. We notice that

$$8 \times 7 \times 6 = \frac{8 \cdot 7 \cdot 6 \cdot 5 \cdot 4 \cdot 3 \cdot 2 \cdot 1}{5 \cdot 4 \cdot 3 \cdot 2 \cdot 1}, \text{ or } \frac{8!}{5!}.$$

Therefore, we can now develop another general rule: if we have n objects, and we are interested in the number of permutations of r of those objects with no repetition allowed, then the total number of permutations is given by $n!/(n-r)!$. The "no repetition allowed" phrase means that, for example, the same person cannot finish in both first and second place. This notation, known as the "r permutations of n," can equivalently be written as $P(n,r) = {}^nP_r = {}_nP_r$—any of these notations works.

Tools Used and Developed

- The number of permutations of n objects is $n!$
- If we have n objects and we are interested in the number of permutations of r of those objects with no repetition allowed, then the total number of permutations is given by

$$\frac{n!}{(n-r)!}.$$

Problem 2.2

Now, let us say our eight runners, Alice, Bob, Cindy, Doug, Evelyn, Frank, George, and Henry, once again designated by {A,B,C,D,E,F,G,H}, run another race, this time for charity. Since this is not a very highly competitive race, it is decided that the top three finishers will each get a race T-shirt showing a picture of the three "champions" standing together. Here, there is no distinction among finishing first, second, or third. How many possible T-shirts are there?

This question is just about how many ways can we pick three people out of eight. Each group of three top finishers, say, (B, G, H), would be one T-shirt, regardless of who was first, second, or third in the race. Runners B, E, G, though, would be a different T-shirt, and so on.

To solve this by brute force, we would systematically list all groups of three, and count them up. It took me quite a while to do this, and several times I made mistakes by duplicating groups of three or leaving out some groups. The groupings are as follows:

{A,B,C},{A,B,D},{A,B,E},{A,B,F},{A,B,G},{A,B,H},{A,C,D},{A,C,E};
{A,C,F},{A,C,G},{A,C,H},{A,D,E},{A,D,F},{A,D,G},{A,D,H},{A,E,F};
{A,E,G},{A,E,H},{A,F,G},{A,F,H},{A,G,H},{B,C,D},{B,C,E},{B,C,F};
{B,C,G},{B,C,H},{B,D,E},{B,D,F},{B,D,G},{B,D,H},{B,E,F},{B,E,G};
{B,E,H},{B,F,G},{B,F,H},{B,G,H},{C,D,E},{C,D,F},{C,D,G},{C,D,H};
{C,E,F},{C,E,G},{C,E,H},{C,F,G},{C,F,H},{C,G,H},{D,E,F},{D,E,G};
{D,E,H},{D,F,G},{D,F,H},{D,G,H},{E,F,G},{E,F,H},{E,G,H},{F,G,H}.

We can count 56 different combinations, or 56 ways to choose three people out of eight.

It really would be nice to have an insight so that we could count more quickly. What we are trying to count here is what is called "combinations" in the language of combinatorics, which means groups where order doesn't matter. For instance, if I am making a bean salad and I choose from three different types of beans, such as lima beans, kidney beans, and fava beans, then that is what my bean salad contains. It doesn't matter if I had said instead that I chose kidney beans, fava beans, and lima beans—it's still the same bean salad.

Insight

The easiest way to understand how to calculate the number of combinations is to start off assuming that order does matter. For example, in our present case, let us return to the prior permutations problem where we have our eight runners, {A,B,C,D,E,F,G,H} and are awarding first-, second-, and third-place medals and order matters. In that case, we calculated that there were 8!/5!, or 336, ways that the medals could be awarded. The key insight is that each group of three runners, say, (B, G, H), can win medals in six different ways, because they can be ordered in 3! different ways:

[(B, G, H), (B, H, G),(G, B, H),(G, H, B),(H, B, G),(H, G, B)].

Therefore, each group of three runners contributes six different permutations to our total of 336. So, there are six times (i.e., 3! times) more permutations than groups of three (i.e., combinations). Thus, to get the total number of groups of three runners; in other words, the number of ways to choose three runners from eight, we would divide 8!/5! by 3!, and get 8!/(5!3!), or 56 different ways to choose three runners out of eight. This calculation is much easier than using the list we made above and then eliminating duplicates. Formally, this is known as calculating the number of combinations without repetition—in other words, each of the runners is different.

Generalizing this, we can now ask how many ways there are to choose r things out of n, often designated by $C(n,r)$. In Problem 1.1, we saw that the "r permutations of n" when order matters is given by

$$P(n,r)=\frac{n!}{(n-r)!}.$$

However, we see that each group of r things contributes $r!$ items to $P(n,r)$. So, we can say that $P(n,r) = r!\ C(n,r)$, or the total number of "r permutations of n," is the number of ways of choosing different groups of r things out of n, or $C(n,r)$, multiplied by the number of ways we can permute each of these r groups, or $r!$.

Therefore, to get $C(n,r)$, we simply divide $P(n,r)$ by $r!$, and we get perhaps the most famous formula in combinatorics:

$$C(n,r)=\frac{n!}{r!(n-r)!}.$$

This is often written as

$$\binom{n}{r}=\frac{n!}{r!(n-r)!},$$

where the left-hand side is read as "n choose r."

We can immediately see that the formula has a certain symmetry: the number of ways of choosing r things out of n is the same as the number of ways of choosing $(n - r)$ things out of n, or in terms of choose notation,

$$\binom{n}{r}=\binom{n}{n-r}=\frac{n!}{r!(n-r)!}.$$

This makes perfect sense: The number of ways of choosing three winners of eight race runners is the same as the number of ways of choosing five losers out of eight runners. For each group of three winners, there is a corresponding group of five losers in the race, with a one-to-one correspondence.

This problem gives us a major tool that we will use over and over again to develop insights into more difficult problems.

Tools Used and Developed

- $P(n,r)=\dfrac{n!}{(n-r)!}$.
- $C(n,r)=\binom{n}{r}=\binom{n}{n-r}=\dfrac{n!}{r!(n-r)!}$.

Problem 2.3

How many arrangements are there of the letters in "roar," where no two *r*s are next to each other?

We know that if "roar" had four distinct letters, like the word "road," there would be 4! = 24 permutations of the letters. However, the two *r*s are indistinguishable, and they can be arranged in 2! different ways, so there are only 4!/2! = 12 permutations of the letters in "roar."

We can list these out:

a o r r	a r o r	a r r o
o a r r	o r a r	o r r a
r a o r	r a r o	r o a r
r o r a	r r a o	r r o a

Of these, we can see that 6 of the 12 arrangements do not have two *r*s next to each other:

a o r r	<u>a r o r</u>	a r r o
o a r r	<u>o r a r</u>	o r r a
<u>r a o r</u>	<u>r a r o</u>	<u>r o a r</u>
<u>r o r a</u>	r r a o	r r o a

This took a little bit of care in listing out the permutations, but it took only a few minutes.

Now let's consider how many arrangements there are for the letters in the word "Mississippi" when no two *i*s are next to each other. We can see right away that our prior direct attack would be a complete nightmare. We need an insight here to help us out in a big way.

Insight

First, let's try to distill the problem to its essence. We have four *i*s and seven other letters, and we want to arrange them so that no two *i*s are next to each other. Therefore, for now, let's not worry about the identity of the other letters, and let's think about just seven generic letters, denoted by *g*, and four *i*s. How would we solve this problem?

For no two *i*s to be next to each other, we would always want a letter between them. Therefore, let's arrange the *g*s with some spacers all around them, denoted by *s*. So we have

$$s\,g\,s\,g\,s\,g\,s\,g\,s\,g\,s\,g\,s\,g\,s.$$

As we see, we have seven *g*s with eight *s*s surrounding them. All we have to do now is to choose four of the *s*s to replace by *i*s, and make the other *s*s disappear, and we would have solved our problem. How many ways are there to choose four *s*s out of eight? That's easy for us now. It's $\binom{8}{4}$, or 70.

However, we don't have seven *g*s; we have the seven other letters in Mississippi, which are four *s*s, two *p*s and one *m*. So, how many ways are there to arrange these letters? If the seven letters were distinct, we could arrange them in 7! ways. However, there are fewer ways here because, for example, we can't tell any of the *s*s apart. These four *s*s, if they were distinguishable, could be arranged in 4! different ways. So, if there are *w* ways to arrange our seven letters (four *s*s, two *p*s, and one *m*), there would be $w \times 4!$ ways to arrange the letters if the *s*s were replaced by four other different letters. Similarly, if the two *p*s were also replaced by two different letters, there would be $w \times 4! \times 2!$ ways to arrange the letters. But, in that case, all of the letters would be different, so this total has to be the 7! ways we could arrange seven different letters. In other words, $w \times 4! \times 2! = 7!$, or $w = 7!/(4!2!) = 105$ ways to arrange the four *s*s, two *p*s, and one *m*.

Now, for each and every one of these 105 arrangements of the four *s*s, two *p*s, and one *m*, we would have 70 ways of putting in four *i*s so that no two *i*s are next to each other, as our analysis above showed. Therefore, there would be a total of $70 \times 105 = 7{,}350$ ways of arranging the letters in "Mississippi" so that there are no two *i*s next to each other. Luckily, we didn't try the old list-and-count approach that we used with "roar." If we had, we would have wasted too many hours trying to count out the possibilities.

Tools Used and Developed

- The choose tool,

$$\binom{n}{k} = \frac{n!}{(n-k)!k!}.$$

- Using spacers to help us count. In general, if we have *n* *g*s and *k* *i*s, and we want to arrange these such that no two *i*s are next to each other, we can do this in $\binom{n+1}{k}$ ways.
- A way to calculate permutations of things not all different. We see that if we have *n* objects, of which some *a*s are alike (such as the four *s*s in Mississippi), and some *b*s are also alike (such as the two *p*s in Mississippi), and some *c*s are alike (such as the four *i*s in Mississippi), and some *d*s are alike (such as the one *m* in Mississippi), and $a + b + c + d = n$, then these can be arranged in $n!/(a!b!c!d!)$ ways.

Problem 2.4

How many positive integer solutions are there for $a + b + c + d = 7$?

We can see that $(a = 2, b = 1, c = 3, d = 1)$ would be one such solution, whereas $(a = 3, b = 1, c = 3, d = 0)$ would not, since not all of the variables are positive integers. Also, in counting up the solutions, we have to keep in mind that ($a = 2$, $b = 1$, $c = 3$, $d = 1$), ($a = 1$, $b = 2$, $c = 3$, $d = 1$), and ($a = 3$, $b = 2$, $c = 1$, $d = 1$), for example, are all different solutions, since they represent different *ordered* quadruples of numbers, although they each correspond to some permutation of the set {1, 1, 2, 3}.

For the purposes of counting, we represent each solution as an ordered quadruple (a, b, c, d) such as (2, 1, 3, 1), wherein we understand that this corresponds to ($a = 2$, $b = 1$, $c = 3$, $d = 1$) written more compactly.

Now we need to list all the solutions, and then count them, trying not to miss anything along the way. We start off with all solutions consisting of three 1s and one 4, then do all solutions with two 1s, one 2, and one 3, and so on, and just try to be systematic:

(1,1,1,4), (1,1,4,1), (1,4,1,1), (4,1,1,1),
(1,1,2,3), (1,1,3,2), (1,2,1,3), (1,2,3,1), (1,3,1,2), (1,3,2,1),
(2,1,1,3), (2,1,3,1), (2,3,1,1), (3,1,1,2), (3,1,2,1), (3,2,1,1),
(2,2,2,1), (2,2,1,2), (2,1,2,2), (1,2,2,2).

That should do it! It actually took a lot of concentration to list these possibilities. For example, for the six quadruples in the second line, we can let $a = 1$, and then list the six permutations of {1,2,3} as the values of (b, c, d). Thinking in this way, we can list the solutions systematically, and count that there are 20 of them.

Seeing that we have made short work of this (about 5 minutes), we decide to do the same calculation, but this time for $a + b + c + d = 14$. We can immediately see that the direct method we used for our first problem would be extremely time consuming, and definitely prone to errors of omission as well as possibly errors of repetition—in other words, we could work for a very long time and still run a high risk of getting the wrong answer. We need some insights.

Insight 1

This insight comes from our study of combinatorics so far, although it is not an easy leap. We can think along the following lines. Since we know the solution has to be in positive integers, we start thinking of the lowest positive integer, 1. We know that to make 14, we need 14 units of 1, but bunched up among the four variables a, b, c, d. Thus, in the ordered quadruple (2, 3, 5, 4), a has been given 2 units of 1, b has been given 3 units of 1, c has been given 5 units of 1, and d has been given 4 units of 1.

Now, let us list these 14 units of 1, designating each with a letter *o*, and put spacers between them, designated by the letter *s*:

o s o s o s o s o s o s o s o s o s o s o s o s o s o.

Now we see that if we eliminate all of the spacers except for three, we will break the *o*s into four groups, each containing some number (at least one) of *o*s. For example,

o o s o o o s o o o o o s o o o o

contains four groups, with 2, 3, 5, and 4 units of 1, respectively. The number of 1s in each group can stand for the values of a, b, c, d, and the above example would correspond to ($a = 2$, $b = 3$, $c = 5$, $d = 4$), or in more compact terms, the ordered quadruple (2, 3, 5, 4) we have been considering above. For any solution we choose, such as (3, 6, 2, 3), we can set up a unique ordering of fourteen *o*s and three *s*s that symbolize that ordered quadruple:

o o o s o o o o o o s o o s o o o.

Thus, we see that there is a one-to-one correspondence between these orderings of 14 *o*s and 3 *s*s and the solutions to our equation $a + b + c + d = 14$ in positive integers. The question then becomes one of ascertaining the number of these orderings. To calculate this number, we notice that between the 14 *o*s are 13 *s*s. To form any ordered quadruple, we must simply choose some 3 of these *s*s to keep, and discard the rest. Thus, we see that the number of solutions is the number of ways we can choose 3 *s*s out of 13, or $\binom{13}{3}$. This is something we learned how to calculate in Problem 2.2, and its value is

$$\frac{13!}{3!10!} = \frac{13 \cdot 12 \cdot 11}{6} = 2 \cdot 13 \cdot 11 = 286.$$

We can test our solution strategy on the first problem, which we solved explicitly by listing and counting the solutions to $a + b + c + d = 7$. Here, we would have 7 *o*s with 6 *s*s in between them, and each solution is generated by choosing 3 *s*s out of the 6 to keep, or $\binom{6}{3}$. This is calculated by

$$\frac{6!}{3!3!} = \frac{6 \cdot 5 \cdot 4}{3!} = 20,$$

which is the answer we got by the direct method.

Insight 2

We can generalize our solution as follows: if we have k variables instead of four, and we wish to count the solutions to the equation $x_1 + x_2 + x_3 + \ldots + x_k = n$ in positive integers, we see that if $k > n$, there can be no solutions, so by assumption, $n \geq k$. Now, we would put n *o*s, and in between them would be $(n - 1)$ *s*s. To break the *o*s into k batches and generate a solution, we need to choose $(k - 1)$ *s*s to keep, so the number of solutions is really the number of ways to choose $(k - 1)$ *s*s out of the $(n - 1)$ we have, given by $\binom{n-1}{k-1}$.

We have solved our problem in a very general way. We see that the solution relied on a tool we already had: how to calculate the number of combinations. The real insight was to cast the problem in terms that allowed us to use this tool, and we did this by using spacers s between the units o, as described above. This was the real insight, and it is a general strategy that we can remember and try to use again and again in yet more creative ways.

Tools Used and Developed

- The choose notation to calculate the number of combinations, that is, how to pick k objects out of n, $\binom{n}{k}$.
- The idea of using spacers s between the units of 1, designated as o, in developing a solution to a linear equation in integers.
- The number of solutions to the equation $x_1 + x_2 + x_3 + \ldots + x_k = n$ in positive integers, given by $\binom{n-1}{k-1}$.

Problem 2.5

How many nonnegative integer solutions are there to $a + b + c + d = 7$?

This problem is related to Problem 2.4. We see here that $(a = 2, b = 1, c = 3, d = 1)$ would be such a solution, as well as $(a = 3, b = 1, c = 3, d = 0)$, because now 0 is an allowed value for the variables. As before, we have to keep in mind that $(a = 3, b = 1, c = 3, d = 0)$ and $(a = 0, b = 1, c = 3, d = 3)$, for example, are different solutions, since they represent different *ordered* quadruples of numbers, even though they each correspond to some permutation of the set $\{3,1,3,0\}$.

In Problem 2.4, we listed all the solutions and then counted them up. This time, we'll try to be a bit smarter. We know that the 20 solutions we found last time for $a + b + c + d = 7$ in positive integers are a subset of the solutions in nonnegative integers, since, as we said above, ($a = 2$, $b = 1$, $c = 3$, $d = 1$) is certainly part of the solutions we are trying to find. To these solutions, we add the following:

- the solutions when one variable is 0,
- the solutions when two variables are 0,
- the solutions when three variables are 0.

When only one variable, say, d, is 0, then we are looking for $a + b + c = 7$ in positive integers. We learned how to do this in Problem 2.4. We discovered that the number of solutions to the equation $x_1 + x_2 + x_3 + \ldots + x_k = n$ in positive integers is given by $\binom{n-1}{k-1}$. Here, $n = 7$ and $k = 3$, so the number of solutions for $a + b + c = 7$ would be $\binom{6}{2} = 6!/(2!4!) = 15$. However, we have to keep in mind that these are the solutions when $d = 0$. But any of the four variables in this first category could have been set to 0, so the total number of solutions when one variable is 0 is $4 \times 15 = 60$. When two variables are 0, say, c and d, this is like looking for $a + b = 7$ in positive integers, which is (1, 6) through (6, 1), or six ways. We could have gotten this from our formula, $\binom{n-1}{k-1}$, which in this case would be $\binom{6}{1} = 6$. However, this is the case when c and d are 0. How about when a and c, or b and d, or any other set of two variables is 0? In fact, there are $\binom{4}{2} = 6$ ways to have two variables out of four be 0, and so in this category, there are another $6 \times 6 = 36$ solutions.

Finally, when three of the variables are 0, then the remaining variable has to be 7, that is, there is only one solution. However, there are four ways of choosing which variable will be 7, so this category adds 4 solutions. Therefore, in addition to our original 20 solutions in positive integers, we add $60 + 36 + 4 = 100$ solutions, for a grand total of 120.

Wow! That's a lot more solutions just by lifting the restriction that all the variables had to be positive integers.

Now we are feeling proud of ourselves that we had the insight to calculate the number of solutions using what we have learned without having to list all the possibilities, so we decide to solve a harder problem: the number of solutions to $a + b + c + d + e = 10$ in nonnegative integers.

We see that the approach we took before, while smart, would be difficult to apply to this problem because we would first have to solve the problem in positive integers, and then consider when one variable is 0, when two variables are 0, when three variables are 0, and when four variables are 0. Also, we see that if we do more difficult problems of the same type, $x_1 + x_2 + x_3 + \ldots + x_k = n$, with higher values of k and of n, our strategy can no longer be practically applied. In short—surprise!—we need an insight.

Insight

In Problem 2.4, we had a great deal of success by using units of 1, which we designated with the letter *o*, and spacers, which we designated by the letter *s*. We see that we can use this same system here: we need ten units of 1, and they can be assigned in any fashion to the five variables a through e. We can use four spacers *s* to make this assignment, but with a particular set of rules, which can be illustrated with the following example:

o s o o s o o o s o o o s o.

Let's look at the number of *o*s to the left of the first *s*, and designate this number as a. Continuing, the number of *o*s between the first and second *s* becomes b, the number of *o*s between the second and third *s* becomes c, the number of *o*s between the third and fourth *s* becomes d, and the number of *o*s to the right of the fourth *s* becomes e. Thus, the above sequence has the assignment ($a = 1$, $b = 2$, $c = 3$, $d = 2$, $e = 1$). With this technique, we can also assign 0 to any or several of the variables. For example, in the sequence *o s o o s o o o s o o o o s*, we see that there is nothing to the right of the fourth *s*, so $e = 0$, and we have ($a = 1$, $b = 2$, $c = 3$, $d = 4$, $e = 0$).

Also, we see that for any variable assignment, we can write an appropriate sequence of four *s*s and ten *o*s to represent that assignment. For instance, for ($a = 3$; $b = 0$; $c = 5$; $d = 0$; $e = 2$), the sequence would be *o o o s s o o o o o s s o o.* We see that since there are no *o*s between the first and second *s*s, $b = 0$, and since there are no *o*s between the third and fourth *s*s, $d = 0$. Therefore, we see that there is a one-to-one correspondence between each sequence of four *s*s and ten *o*s, and each variable assignment for $a + b + c + d + e = 10$ in nonnegative integers. Finding the number of these sequences (and hence the number of solutions to the linear equation in nonnegative integers) is the same as simply finding how to pick 4 spaces for the *s*s out of the 14 available spaces, or $\binom{14}{4} = \binom{14}{10}$, which is calculated as

$$\binom{14}{10} = \frac{14!}{4!10!} = \frac{14 \cdot 13 \cdot 12 \cdot 11}{4!} = 7 \cdot 13 \cdot 11 = 1{,}001.$$

By using the same reasoning, we can say that, for $x_1 + x_2 + x_3 + \ldots + x_k = n$ in general, there will be $\binom{n+k-1}{k-1} = \binom{n+k-1}{n}$ solutions in nonnegative integers. We can apply this solution to the first, easier problem for which $a + b + c + d = 7$. We see that the number of solutions is $\binom{10}{3} = 10!/(3!7!) = 120$, which is the same answer we got by using brute force. It is amazing how much effort an insight can save.

Extension

The insight that we just discovered can come in handy in solving a variety of problems that may not initially seem to be related to it. For example, suppose we want to find the number of integers between 1 and 1,000 whose digits sum to 8. The natural way to proceed is probably to think of single-digit numbers first, and here we have just 8. Then, for two-digit numbers, we would have 17, 71, 26, 62, 35, 53, 44, and 80. So far so good. Now, things get complex. We would need to take a number, say, 17, and break the 7 down into all of its possibilities, such as 116, 161, 125, 152, and so on. What if we did the same for all the others? If we take 71, and break the 7 down now, we could get 611 (which we didn't get before) but we could also get 116, which we generated previously. We can see that this will be a hopeless mess, with overcounting, duplications, and other pitfalls. It will be a real challenge to keep track of the situation.

However, an additional insight, along with the counting technique we just developed, rescues us from this mess. First, let us exclude 1,000, since its digits don't sum to 8. Now let us assume that we are searching through 1–999, and turn all of the numbers into three-digit numbers. We can do that by putting zeroes appropriately to the left of one- or two-digit numbers. For example, 6 can be thought of as 006, 23 can become 023, and so forth. If we let our three digits be x_1, x_2, and x_3, we can interpret the problem as finding the number of solutions of $x_1 + x_2, + x_3 = 8$ in nonnegative integers. Each solution will correspond to one of the integers between 1 and 999. Since the required sum is 8, we know that each of x_1, x_2, and x_3 is a single digit, since the largest any of them can be is 8, so this approach will work fine. Therefore, the required answer is $\binom{10}{2} = 10!/(8!2!) = 45$. If we had been given this problem initially, we may not have seen that it is directly convertible to calculating the number of solutions of a linear equation in nonnegative integers. It is important to develop these sorts of links, as it is unlikely that someone will say, "Hey, kid, how many solutions are there in nonnegative integers for

$x_1 + x_2, + x_3 = 8$?" They are much more likely to give the question a bit of twist, like this extension problem.

Tools Used and Developed

- The choose notation to calculate the number of combinations, or how to pick k objects out of n, $\binom{n}{k}$.
- The idea of again using spacers, designated as s, between units of 1, designated as o, in the number of solutions to a linear equation in non-negative integers.
- The number of solutions to the equation $x_1 + x_2 + x_3 + \ldots + x_k = n$ in nonnegative integers is given by $\binom{n+k-1}{k-1}$.

Problem 2.6

What is the coefficient of x^3y when we expand $(x + y)^4$?

This solution is not so bad. We know how to square a binomial: $(x + y)^2 = x^2 + 2xy + y^2$. We can now multiply this by $(x + y)$ and, after some algebra, get $(x + y)^3 = x^3 + 3x^2 y + 3xy^2 + y^3$. We multiply again by $(x + y)$ to get $(x + y)^4$ in the following way. First, we multiply all the terms by x, and get $x^4 + 3x^3 y + 3x^2 y^2 + xy^3$. Then we multiply all the terms by y to get $x^3 y + 3x^2 y^2 + 3xy^3 + y^4$. Adding these two results, we get: $x^4 + 4x^3 y + 6x^2 y^2 + 4xy^3 + y^4$. Therefore, the coefficient of x^3y is 4.

But what if we want to determine the coefficient of x^3y^6 in $(x + y)^9$? We can see that this will quickly become very tedious, and we need an insight to do this kind of problem, especially as the powers become larger and larger.

Insight

The insight here comes from what we've done so far about combinations, and leads to a very interesting structure in combinatorics known as Pascal's triangle. To start, let's look at the expansion of $(x + y)^9$. We can write this in longhand as

$$(x+y)\,(x+y)\,(x+y)\,(x+y)\,(x+y)\,(x+y)\,(x+y)\,(x+y)\,(x+y).$$

If we multiply this out, we see that the resulting terms will, in one way or another, be some combination that results from taking either an x or a y from each binomial. Since there are 9 binomials, and from each one we can choose either an x or a y (2 choices), we will have 2^9 terms. In each of these terms, the sum of the powers of x and y will have to be 9. Also, there are only 10 types of terms, ranging from x^9y^0 to x^0y^9, so many of the 2^9 terms will be duplicates. Let's focus on one specific term, such as x^3x^6. We can see that this term can come about in many different ways. For example, we can choose xs from the first three binomials and ys from the remaining six to get x^3y^6:

$$\underset{\uparrow}{(x+y)}\underset{\uparrow}{(x+y)}\underset{\uparrow}{(x+y)}(x+y)(x+y)(x+y)(x+y)(x+y)(x+y)$$

Alternatively, we can choose xs from the second, sixth, and ninth binomials, and ys from the rest to also get x^3y^6:

$$(x+y)\underset{\uparrow}{(x+y)}(x+y)(x+y)(x+y)\underset{\uparrow}{(x+y)}(x+y)(x+y)\underset{\uparrow}{(x+y)}$$

Therefore, we see that we can get x^3y^6 in many different ways. The question, of course, is how many? It is precisely the number of ways that we can choose 3 binomials out of 9 to contribute the x term. We already know how to calculate this. It is simply "9 choose 3," or $\binom{9}{3} = 9!/(3!6!)$. This comes out to 84, so the coefficient of x^3y^6 in the expansion of $(x + y)^9$ is 84.

This insight now gives us the general rule for expanding binomials. When we have $(x + y)^n$, we see that for any term $x^{n-k}y^k$, its coefficient will be $\binom{n}{n-k} = \binom{n}{k}$. Therefore, we can write:

$$(x+y)^n = \binom{n}{0}x^n + \binom{n}{1}x^{n-1}y^1 + \binom{n}{2}x^{n-2}y^2 + \cdots + \binom{n}{n}y^n,$$

remembering that x^0 and y^0 are both equal to 1, and that $\binom{n}{0} = \binom{n}{n} = 1$.

Each coefficient of a term in the binomial expansion is known as a binomial coefficient. If we look at these coefficients as we expand the binomial $(x + y)^n$ to higher and higher powers of n, we can line them up on top of each other, and we get Pascal's triangle, one of the most interesting objects in math. Let's do that:

$$(x+y)^0 = 1,$$

$$(x+y)^1 = 1x + 1y,$$

$$(x+y)^2 = 1x^2 + 2xy + 1y^2,$$

$$(x+y)^3 = 1x^3 + 3x^2y + 3xy^2 + 1y^3,$$

$$\vdots$$

$$(x+y)^n = \binom{n}{0}x^n + \binom{n}{1}x^{n-1}y^1 + \binom{n}{2}x^{n-2}y^2 + \cdots + \binom{n}{n}y^n.$$

If we now take all of the coefficients, and make one row for each power, we get Pascal's triangle; below are the first six rows, row 0 through row 5, as an illustration. The triangle continues on infinitely.

$$\begin{array}{ccccccccccc}
 & & & & & 1 & & & & & \\
 & & & & 1 & & 1 & & & & \\
 & & & 1 & & 2 & & 1 & & & \\
 & & 1 & & 3 & & 3 & & 1 & & \\
 & 1 & & 4 & & 6 & & 4 & & 1 & \\
1 & & 5 & & 10 & & 10 & & 5 & & 1
\end{array}$$

So each row in Pascal's triangle represents the coefficients of the terms in the binomial expansion of $(x + y)^n$, beginning with "row 0". It's a bit odd to have the first row called "row 0," but this way, for example, row 3 (which is actually the *fourth* row down, since we have rows 0, 1, 2, and then row 3) consists of the numbers 1, 3, 3, 1. These are precisely the coefficients of the terms of $(x + y)^3 = 1x^3 + 3x^2 y + 3xy^2 + 1y^3$, proceeding from the highest to the lowest power of x. We can thus think of the numbers in row n of Pascal's triangle as the coefficients of the terms in the expansion of $(x + y)^n$; in other words, $\binom{n}{0}\binom{n}{1}\binom{n}{2}\cdots\binom{n}{n}$. Notice that row n of Pascal's triangle will have $(n + 1)$ numbers in it, with the first and last numbers being 1.

There are almost endless numbers of interesting properties of Pascal's triangle, but we'll leave those for later. It is interesting, though, to meet Pascal's triangle here, and to see its origin. We can generate the rows of Pascal's triangle without doing all of the multiplication—that is, without expanding $(x + y)^n$ to see what the coefficients are. Notice that the first and last number in every row is 1. In between, each number is just the sum of the two numbers immediately above it, just to the left and just to the right. For example, look at the fourth and fifth rows of Pascal's triangle:

```
     1   4   6   4   1
      \ / \ / \ / \ /
  1    5   10  10   5    1
```

The number 5 is the sum of the numbers 1 and 4 just above it, and the number 10 is the sum of the numbers 4 and 6 just above it, and so on. We'll soon see how this particular property plays into the solution to Problem 2.7.

Tools Used and Developed

- The choose tool,

$$\binom{n}{k} = \frac{n!}{(n-k)!k!}.$$

- The rules for the coefficients of a binomial expansion:

$$(x+y)^n = \binom{n}{0}x^n + \binom{n}{1}x^{n-1}y^1 + \binom{n}{2}x^{n-2}y^2 + \cdots + \binom{n}{n}y^n.$$

- Pascal's triangle.

Problem 2.7

Calculate the following sum, now that we know about binomial coefficients:

$$\binom{7}{0}+\binom{7}{1}+\binom{7}{2}+\binom{7}{3}+\binom{7}{4}+\binom{7}{5}+\binom{7}{6}+\binom{7}{7}.$$

We smile and shrug. This is easy. We know how to calculate a binomial coefficient:

$$\binom{n}{k} = \frac{n!}{(n-k)!k!}.$$

Moreover, we can even be a bit tricky, since we also know that $\binom{n}{n-k} = \binom{n}{k}$. Therefore, for example, $\binom{7}{2} = \binom{7}{5}$. Thus, we can reduce our problem to the following:

$$2\binom{7}{0} + 2\binom{7}{1} + 2\binom{7}{2} + 2\binom{7}{3},$$

and we can calculate each of the four terms:

$$\binom{7}{0} = \frac{7!}{0!\,7!} = 1,$$

$$\binom{7}{1} = \frac{7!}{1!6!} = \frac{7\cdot(6!)}{1!\,6!} = 7,$$

$$\binom{7}{2} = \frac{7!}{2!\,5!} = \frac{7\cdot 6\cdot(5!)}{2!\,5!} = 21,$$

$$\binom{7}{3} = \frac{7!}{3!\,4!} = \frac{7\cdot 6\cdot 5.(4!)}{3!\,4!} = 35.$$

So our answer is 2(1) + 2(7) +2(21) +2(35) = 128.

That wasn't bad. It only took a few minutes. But now we want to do the following sum:

$$\binom{11}{0} + \binom{11}{1} + \cdots + \binom{11}{11}.$$

We see that this could now start to take a while. Some insight would certainly help us cut down on the tedium factor.

Insight 1

Let's work with our first problem, since we already know the answer from above. Our first insight is to relate the problem to what we just learned about Pascal's triangle. We know that the numbers we need in our sum, $\binom{7}{0} + \ldots + \binom{7}{7}$, are the numbers that make up row 7 (the eighth row) of Pascal's triangle. Also, we know a fast way to generate each row from the one before. Problem 2.6 showed Pascal's triangle through row 5, so we just use that to quickly generate the next two rows:

```
                1
              1   1
            1   2   1
          1   3   3   1
        1   4   6   4   1
      1   5  10  10   5   1
       \ / \ / \ / \ / \ /
    1   6  15  20  15   6   1
     \ / \ / \ / \ / \ / \ /
  1   7  21  35  35  21   7   1
```

If we add the numbers in the last row, we get 128. This was a bit quicker than actually calculating the binomial coefficients, but not by that much. Using this method to solve $\binom{n}{0} + \binom{n}{1} + \ldots + \binom{n}{n}$ for larger n would get to be tiresome, as we would need to keep calculating more and more rows of Pascal's triangle. Do we have a better insight?

Insight 2

We can do our calculation much more efficiently by remembering what the binomial coefficients (and the numbers in any row of Pascal's triangle) represent. The numbers in row n represent the coefficients of the terms in the expansion of $(x + y)^n$, such that

$$(x+y)^n = \binom{n}{0}x^n + \binom{n}{1}x^{n-1}y^1 + \binom{n}{2}x^{n-2}y^2 + \cdots + \binom{n}{n}y^n.$$

Here, x and y are just variables, which can take on any values and the above expression will still be true. Do we have the insight yet? Well, how about if we just let x and y both equal 1? In that case, all terms of the form $x^{n-k}y^k$ will equal 1, and the expression reduces to

$$(1+1)^n = \binom{n}{0} + \binom{n}{1} + \cdots + \binom{n}{n}.$$

In other words, we have just shown that the sum of the binomial coefficients in row n of Pascal's triangle is just 2^n. This is a one-step solution. When $n = 7$, as in our first problem, the answer is just $2^7 = 128$. For our harder problem, with $n = 11$, we can say without any messy calculations that the answer is $2^{11} = 2{,}048$. It doesn't get much slicker than that. I must admit that when I first did this problem, I used a method like the first insight, and when I saw the solution in the second insight, I was just blown away. I kept asking, "Wow, why didn't I think of that?" That, to me, is the most fun part of doing math.

Insight 3

Now that we've solved our problem in a very slick way, let's step back and take a broader view, as it will lead us into a problem-solving approach that some people refer to as "combinatorial reasoning." In other words, let's solve the problem by assuming that we are counting something, and find two different ways to count the same thing. Both of the answers have to be equal, assuming that our counting is correct. Let us get back to the basic meaning of $\binom{7}{2}$. It counts the number of ways of picking 2 objects out of 7. For example, if we are going to Disneyland, and we have 7 people that might want to go, there are $\binom{7}{2}$ ways of picking just 2 of them, that is, forming groups of 2. Similarly, $\binom{7}{3}$ is the number of ways of picking 3 people to go, and so on. Now let us think about adding up all of these ways.

If we add up all the groups that contain 0 people (only one group here), 1 person, 2 people, 3 people, and so on, through the group that contains all 7 people, we see that none of these groups overlap. They are all different groups of people we could take to Disneyland. Also, adding up all of the groups covers all of the possible ways we could take 7 people to Disneyland. This would be given by our original problem,

$$\binom{7}{0} + \binom{7}{1} + \binom{7}{2} + \binom{7}{3} + \binom{7}{4} + \binom{7}{5} + \binom{7}{6} + \binom{7}{7}.$$

Now let's count up this same total in a different way. We have 7 people to choose from, and for each person, there are 2 possibilities when we are taking people: either they go, or they don't. So the first person generates 2 possibilities, the second person generates 2 possibilities, and so on. Altogether, the 7 people generate $2 \times 2 \times 2 \times 2 \times 2 \times 2 \times 2 = 2^7 = 128$ ways of taking people to Disneyland. Thus, using combinatorial reasoning and generalizing our approach, we again see that

$$\binom{n}{0}+\binom{n}{1}+\cdots+\binom{n}{n}=2^n.$$

Tools Used and Developed

- Pascal's triangle.
- $(x+y)^n=\binom{n}{0}x^n+\binom{n}{1}x^{n-1}y^1+\binom{n}{2}x^{n-2}y^2+\cdots+\binom{n}{n}y^n.$
- A practical use of the binomial expansion to get the identity

$$\binom{n}{0}+\binom{n}{1}+\cdots+\binom{n}{n}=2^n.$$

- An introduction to the idea of combinatorial reasoning—solving a problem by finding two ways of counting the same thing.

Problem 2.8

Calculate the following sum, using what we have learned about the sum of the binomial coefficients:

$$\binom{8}{0}+\binom{8}{2}+\binom{8}{4}+\binom{8}{6}+\binom{8}{8}.$$

Although we know from our last problem that $\binom{n}{0}+\binom{n}{1}+\ldots+\binom{n}{n}=2^n$, we don't know how to handle the sum of just even "choices." As we stare at the problem for a while, we figure it's just easier to get the answer directly, so we pull out our pencils and roll. We know how to calculate a binomial coefficient,

$$\binom{n}{k} = \frac{n!}{(n-k)!k!},$$

and we can use the fact that $\binom{n}{n-k} = \binom{n}{k}$, so, for example, $\binom{8}{2} = \binom{8}{6}$. Thus, we can reduce this problem to

$$2\binom{8}{0} + 2\binom{8}{2} + \binom{8}{4}.$$

We now calculate each of the terms:

$$\binom{8}{0} = \frac{8!}{0!\,8!} = 1,$$

$$\binom{8}{2} = \frac{8!}{2!6!} = \frac{8\cdot 7\cdot (6!)}{2!\,6!} = 28,$$

$$\binom{8}{4} = \frac{8!}{4!\,4!} = \frac{8\cdot 7\cdot 6\cdot 5\cdot (4!)}{4!\,4!} = 70,$$

and our answer is 2(1) + 2(28) + 70 = 128.

That wasn't bad. It only took a few minutes. So now we think we can do the following sum:

$$\binom{16}{0} + \binom{16}{2} + \binom{16}{4} + \cdots + \binom{16}{16}.$$

We see, though, that this will be a serious problem if we do the calculation directly. We need an insight.

Insight

We need to be able to use information we've learned to come to new insights. In Problem 2.7, we used the expansion

$$(x+y)^n = \binom{n}{0}x^n + \binom{n}{1}x^{n-1}y^1 + \binom{n}{2}x^{n-2}y^2 + \cdots + \binom{n}{n}y^n,$$

and we substituted $x = 1$ and $y = 1$ to calculate the sum of the binomial coefficients:

$$(1+1)^n = 2^n = \binom{n}{0}+\binom{n}{1}+\cdots+\binom{n}{n}. \tag{2.1}$$

However, the binomial expansion is perfectly general, and we could alternatively substitute $x = 1$ and $y = -1$. In this case, all of the odd powers of y become negative, and we get:

$$(1+-1)^n = \binom{n}{0}-\binom{n}{1}+\binom{n}{2}-\binom{n}{3}+\cdots+(-1)^n\binom{n}{n}=0. \tag{2.2}$$

Thus, we see that the sum of the binomial coefficients with alternating signs equals 0. If we add Equations (2.1) and (2.2):

$$2^n = \binom{n}{0}+\binom{n}{1}+\binom{n}{2}+\binom{n}{3}\cdots+\binom{n}{n},$$

$$0 = \binom{n}{0}-\binom{n}{1}+\binom{n}{2}-\binom{n}{3}\cdots+\binom{n}{n},$$

we see that all of the odd terms cancel, and we get, for even n,

$$2^n = 2\binom{n}{0}+2\binom{n}{2}+2\binom{n}{4}+\cdots+2\binom{n}{n}.$$

Therefore, we have

$$\frac{2^n}{2} = 2^{n-1} = \binom{n}{0}+\binom{n}{2}+\binom{n}{4}+\cdots+\binom{n}{n}.$$

We could have arrived at the same conclusion even more directly by noticing that the sum of the binomial coefficients with alternating signs is 0, meaning that, if n is even,

$$\binom{n}{0}+\binom{n}{2}+\binom{n}{4}+\cdots+\binom{n}{n}=\binom{n}{1}+\binom{n}{3}+\binom{n}{5}+\cdots+\binom{n}{n-1}.$$

Since both of these sets of coefficients together sum to 2^n, and the sum on the left equals the sum on the right, we see that each sum tallies to $2^n/2 = 2^{n-1}$. Thus, as a bonus, we also see that the sum of all the binomial coefficients with an odd bottom number is 2^{n-1} as well.

Going back to our first problem, $\binom{8}{0}+\binom{8}{2}+\binom{8}{4}+\binom{8}{6}+\binom{8}{8}$, we see that $n = 8$, so the sum is $2^7 = 128$, as we had obtained directly. For our bigger problem $\binom{16}{0}+\binom{16}{2}+\binom{16}{4}+\ldots+\binom{16}{16}$, we know the sum will be $2^{15} = 32{,}768$.

So we see how an insight from one problem directly allowed us to gain an insight into solving a related problem. That is exactly how math builds on itself, and we see that insights are probably more about learning and practice than raw intelligence (although that certainly doesn't hurt!).

Tools Used and Developed

- $(x+y)^n=\binom{n}{0}x^n+\binom{n}{1}x^{n-1}y^1+\binom{n}{2}x^{n-2}y^2+\cdots+\binom{n}{n}y^n,$

  from which we derived that

  $$\binom{n}{0}+\binom{n}{1}+\cdots+\binom{n}{n}=2^n.$$

- $\binom{n}{0}-\binom{n}{1}+\binom{n}{2}-\binom{n}{3}+\cdots+(-1)^n\binom{n}{n}=0.$

- $\binom{n}{0}+\binom{n}{2}+\binom{n}{4}+\cdots+\binom{n}{2k}=2^{n-1}.$

Problem 2.9

Calculate the following sum:

$$\binom{8}{0}^2+\binom{8}{1}^2+\binom{8}{2}^2+\cdots+\binom{8}{8}^2.$$

Although we've done some problems with binomial coefficients, and we know from Problem 2.7 that $\binom{n}{0} + \binom{n}{1} + \ldots + \binom{n}{n} = 2^n$, we don't know how to handle the sum of the *squares* of the binomial coefficients. If we try

$$\left[\binom{n}{0}+\binom{n}{1}+\cdots+\binom{n}{n}\right]^2,$$

we would get the terms we want, but we would also get all the cross terms of the form $\binom{n}{i}\binom{n}{j}$, which we have no easy way to eliminate. Once again, it seems that it just may be easier to get the answer directly, simplifying a bit with the fact that $\binom{n}{n-k} = \binom{n}{k}$. Thus, we can reduce the problem to

$$2\binom{8}{0}^2+2\binom{8}{1}^2+2\binom{8}{2}^2+2\binom{8}{3}^2+\binom{8}{4}^2.$$

We now calculate each of the terms:

$$\binom{8}{0}=\frac{8!}{0!\,8!}=1,$$

$$\binom{8}{1}=\frac{8!}{1!\,7!}=8,$$

$$\binom{8}{2}=\frac{8!}{2!6!}=\frac{8\cdot 7\cdot(6!)}{2!\,6!}=28,$$

$$\binom{8}{3}=\frac{8!}{3!5!}=\frac{8\cdot 7\cdot 6\cdot(5!)}{3!\,5!}=56,$$

$$\binom{8}{4}=\frac{8!}{4!\,4!}=\frac{8\cdot 7\cdot 6\cdot 5\cdot(4!)}{4!\,4!}=70.$$

From these coefficients, our answer is $2(1)^2 + 2(8)^2 + 2(28)^2 + 2(56)^2 + 2(70)^2 = 12{,}870$. That wasn't so terrible.

Spurred on, we try to do it for 16:

$$\binom{16}{0}^2+\binom{16}{1}^2+\binom{16}{2}^2+\cdots+\binom{16}{16}^2.$$

We see that could go on all day, so we need to solve the problem in general. Therefore, we need an insight.

Insight

In the last four problems, we used the binomial expansion,

$$(x+y)^n=\binom{n}{0}x^n+\binom{n}{1}x^{n-1}y^1+\binom{n}{2}x^{n-2}y^2+\cdots+\binom{n}{n}y^n,$$

and had great success by substituting clever values for x and y. However, we don't see how we can do that here. Therefore, we try a second strategy that we learned in Problem 2.7, when we summed the binomial coefficients: counting things in two different ways. For example, when we counted all of the different ways we could take none, some, or all of n people to Disneyland, we realized that we had found a good way to sum the binomial coefficients and showed that

$$\sum_{k=0}^{n}\binom{n}{k}=2^n.$$

Based on this, we decide to write our sum a bit differently, once again using our wonderful identity $\binom{n}{n-k}=\binom{n}{k}$. Therefore, we can write:

$$\binom{n}{k}^2=\binom{n}{k}\binom{n}{n-k}$$

for each term, so our sum is now

$$\binom{n}{0}\binom{n}{n}+\binom{n}{1}\binom{n}{n-1}+\binom{n}{2}\binom{n}{n-2}+\cdots+\binom{n}{n}\binom{n}{0}.$$

Writing the sum in this way, all of a sudden, it hits us! We have an insight about how we can count the sum of the squares of the binomial coefficients.

Imagine that our class has $2n$ students, n boys and n girls. We need to pick half the class, or n students (with gender not being an issue), to try a new math book. If the group trying the new math book consists of k boys, then there are $\binom{n}{k}$ ways of picking them. Also, we would need $n - k$ girls (for a total of n students), and there would be $\binom{n}{n-k}$ ways of picking those girls. Thus, there are $\binom{n}{k}\binom{n}{n-k}$ ways of picking k boys and $n - k$ girls to try the new book. Here, k is a dummy variable. For example, if $k = 3$, there would be $\binom{n}{3}\binom{n}{n-3}$ total ways to pick 3 boys and $n - 3$ girls. However, there is nothing special about $k = 3$. We could instead have chosen 4 boys and $n - 4$ girls, or 5 boys and $n - 5$ girls, and so on. Thus, to get the total number of ways to pick n students, we have to let k range all the way from 0 boys to n boys, and find the sum,

$$\sum_{k=0}^{n}\binom{n}{k}\binom{n}{n-k},$$

but because $\binom{n}{n-k} = \binom{n}{k}$, this sum is equivalent to

$$\sum_{k=0}^{n}\binom{n}{k}^2.$$

However, there is an easier way to get the total number of ways of picking the n students we need. Since gender doesn't matter, and we want to count all possibilities, we simply want all the ways to pick n students out of the $2n$ total in the class, or $\binom{2n}{n}$. These two totals have to match, so we have established that

$$\sum_{k=0}^{n}\binom{n}{k}^2 = \binom{2n}{n}.$$

If this is correct, then

$$\binom{8}{0}^2 + \binom{8}{1}^2 + \binom{8}{2}^2 + \cdots + \binom{8}{8}^2 = \binom{16}{8}.$$

When we check this directly, we get $\binom{16}{8} = 16!/(8!8!) = 12{,}870$.

To calculate the bigger problem,

$$\binom{16}{0}^2+\binom{16}{1}^2+\binom{16}{2}^2+\cdots+\binom{16}{16}^2,$$

we need to find $\binom{32}{16}$ = 32!/(16!16!). This takes a bit of work to calculate directly, but it gives us 601,080,390, and in a fashion much faster than calculating each of the binomial coefficients separately, squaring each of them, and then adding the results.

Tools Used and Developed

- Counting the same thing in two different ways.
- The identity $\binom{n}{n-k} = \binom{n}{k}$, leading to the idea that

$$\binom{n}{k}^2=\binom{n}{k}\binom{n}{n-k}.$$

- $\sum_{k=0}^{n}\binom{n}{k}^2=\binom{2n}{n}.$

Problem 2.10

How many integers between 1 and 100 inclusive are not divisible by 3, 5, or 7?

How can we find this? We can list all of the numbers and then delete all that are divisible by 3, 5, or 7, and count the rest.

1	2	3	4	5	6	7	8	9	10
11	12	13	14	15	16	17	18	19	20
21	22	23	24	25	26	27	28	29	30
31	32	33	34	35	36	37	38	39	40
41	42	43	44	45	46	47	48	49	50
51	52	53	54	55	56	57	58	59	60
61	62	63	64	65	66	67	68	69	70
71	72	73	74	75	76	77	78	79	80
81	82	83	84	85	86	87	88	89	90
91	92	93	94	95	96	97	98	99	100

Taking out all of the numbers divisible by 3 leaves us with:

1	2		4	5		7	8		10
11		13	14		16	17		19	20
	22	23		25	26		28	29	
31	32		34	35		37	38		40
41		43	44		46	47		49	50
	52	53		55	56		58	59	
61	62		64	65		67	68		70
71		73	74		76	77		79	80
	82	83		85	86		88	89	
91	92		94	95		97	98		100

Now taking out all of the numbers that end in 5, since we know that those are multiples of 5, leaves us with:

1	2		4		7	8	
11		13	14	16	17		19
	22	23		26		28	29
31	32		34		37	38	
41		43	44	46	47		49
	52	53		56		58	59
61	62		64		67	68	
71		73	74	76	77		79
	82	83		86		88	89
91	92		94		97	98	

Finally, we physically sort through the numbers and remove the remaining multiples of 7:

1	2		4			8	
11		13		16	17		19
	22	23		26			29
31	32		34		37	38	
41		43	44	46	47		
	52	53				58	59
61	62		64		67	68	
71		73	74	76			79
	82	83		86		88	89
	92		94		97		

We are left with 45 numbers—the numbers that are not divisible by 3, 5, or 7. Although this was a tedious process, it was possible to do it by hand. But what if we wanted to do this for 1–1,000? We must find a better system, or an insight.

Insight

For the first part, considering the numbers 1–100, every third number will be divisible by 3; hence, to figure out how many of these "third numbers," so to speak, exist in 1–100, we calculate: $100/3 = 33\frac{1}{3}$. This means that from 1 to 100 there are 33 groups of three numbers (i.e., 1, 2, 3 is one group; 4, 5, 6 is a second group; and so on until 97, 98, 99 is our last group of three in this set of numbers). This means that we have 33 of these groups, with the remaining $\frac{1}{3}$ being the number 100, which did not fit into a group of three. Hence, we can subtract the 33 multiples of 3 from our 100 numbers, and we have 67 remaining numbers.

We can easily follow the same logic for 5. Since every fifth number is divisible by 5, there are $100/5 = 20$ numbers that are divisible by 5. Hence, we can subtract away the 20 multiples of 5 from the remaining numbers, leaving $67 - 20 = 47$ numbers.

Finally, we can perform the same operation to find the multiples of 7: $100/7 = 14\frac{2}{7}$. Just as with the multiples of 3, this means that there are 14 multiples of 7 in the numbers 1–100. Hence, we can subtract these multiples from our remaining numbers, leaving $47 - 14 = 33$ numbers.

But what about numbers such as 15, which is a multiple of both 3 and 5? We have subtracted the number 15 out of our set of 100 numbers twice, once when we subtracted out the multiples of 3 and a second time when we subtracted out the multiples of 5. We have done this with all multiples of 3 and 5, multiples of 3 and 7, and multiples of 5 and 7. To correct for this error, we must add these multiples back in a single time.

Let us add back in all multiples of 15 (numbers that are both multiples of 3 and 5). How many of these are there? This is easy enough to figure out. We will just follow the same pattern we were using before, or $100/15 = 6\frac{2}{3}$. This means that there are 6 multiples of 15 in the numbers 1–100, so we add back these 6 multiples to the 33 numbers we had remaining, and now we have 39 numbers.

Let's now figure out how many multiples of 21 there are: $100/21 = 4\frac{16}{21}$. This means that there are 4 multiples of 21 in the numbers 1–100, and we add these back to our 39 numbers to get 43 remaining numbers.

Now let's figure out how many multiples of 35 there are: $100/35 = 2\frac{6}{7}$. This means that there are 2 multiples of 35 in the numbers 1–100, and we add these back to our 43 numbers to get 45 remaining numbers, the same as we got by counting.

We have now eliminated all of the numbers that are multiples of 3, 5, and 7 while conscientiously making sure to eliminate each of these numbers only once. This method, known as the inclusion-exclusion principle, gives us the correct amount of numbers remaining.

Now let's do the same problem for the numbers 1–1,000. We can use the inclusion-exclusion principle to eliminate any multiples of 3, 5, and 7. We will use relatively the same method, but we will work a bit faster and add an extra step toward the end.

The number of multiples of 3 in the numbers 1–1,000 are 1,000/3= $333\frac{1}{3}$, so, 1,000 − 333 = 667 numbers are remaining. Let's now eliminate the multiples of 5: 1,000/5= 200, so, 667 – 200 = 467 numbers are remaining. Finally, let's eliminate the multiples of 7: 1,000/7 = $142\frac{6}{7}$, and 467 – 142 = 325 numbers are remaining.

Just as before, we must now add back in the multiples of 15 (multiples of both 3 and 5), 21 (multiples of both 3 and 7), and 35 (multiples of both 5 and 7) because we have eliminated them twice in our count. For multiples of 15, there are 1,000/15= $66\frac{2}{3}$, or 66 numbers. Similarly, for 21, there are 1,000/21 = $47\frac{13}{21}$, or 47 numbers; and finally, for 35, there are 1,000/35 = $28\frac{4}{7}$, or 28 numbers. When we add back all of these duplicates, we have 325 + (66 + 47 + 28) = 466 numbers remaining.

But we're not done yet. We have now eliminated the multiples of 3, 5, and 7, and then added back the multiples of 15, 21, and 35, but what do we do about numbers such as 105, which are multiples of 3, 5, and 7? Let's think about what has happened to the number 105 throughout our inclusion-exclusion process. We would have counted out 105 when we eliminated the multiples of 3, and then counted it out again when we eliminated the multiples of 5, and then counted it out a third time when we eliminated the multiples of 7. So, in total, we eliminated the number 105 a total of three times. To counteract this, though, we counted it back in once when we added back the multiples of 15, we counted it back in again when we added back the multiples of 21, and we counted it back in a third time when we added back the multiples of 35. So, in total, we added back 105 a total of three times as well. This means that when discounting multiples of 3, 5, and 7, we discounted 105 a total of three times, but then added it back into the count a total of three times, essentially undoing any elimination.

So, what should we do now? We know that numbers that are multiples 105 (210, 315, ...) should be eliminated, but obviously only once, as we would cross them out of a table of numbers a single time. Thus, at this point, we must discount the multiples of 105, which are multiples of 3, 5, and 7, one final time. But how many multiples of 105 are there in the numbers 1–1,000? We find this by calculating $1{,}000/105 = 9\frac{11}{21}$, so there are 9 of these multiples. Therefore, our final answer is that there are $466 - 9 = 457$ numbers remaining.

Indeed, if we were to write out all of the numbers 1–1,000 and cross out any number that was a multiple of 3, 5, or 7, we would have 457 numbers remaining. The inclusion-exclusion principle makes calculations like these easy and elegant!

Tools Used and Developed

- Inclusion-exclusion principle.

Problem 2.11

How many ways are there to make change for a quarter using only dimes, nickels, and pennies?

For this problem, it is not difficult to simply list all of the 12 possibilities (as shown in the following table).

Dimes	*Nickels*	*Pennies*
2	1	0
	0	5
1	3	0
	2	5
	1	10
	0	15
0	5	0
	4	5
	3	10
	2	15
	1	20
	0	25

But what if we wanted to know how many ways there are to change a dollar, using half-dollar coins, quarters, dimes, nickels, and pennies? We know we can solve this problem in the same way: brute force. We will have to be very careful, however, not to lose track of any possibilities, so we have to be very systematic. The brute force approach for this problem is shown below. It takes into account that once we have decided the number of half-dollars, quarters, dimes and nickels, the number of pennies is, of course, decided by the number of nickels, so the pennies don't have to be listed separately and are included in the "nickels" column.

If we look at the third line in the table, for example, we use 1 half-dollar, 1 quarter, and 2 dimes, for a total of 95 cents. This combination leads to two remaining possibilities for the nickel column, using either 1 nickel, or 0 nickels and 5 pennies. The "number of possibilities" column is just a count of the possibilities for nickels and pennies, given that the half-dollars, quarters, and dimes are as shown.

Half-dollars	*Quarters*	*Dimes*	*Nickels*	*Number of Possibilities*
2	0	0	0	1
1	2	0	0	1
1	1	2	(0–1)	2
1	1	1	(0–3)	4
1	1	0	(0–5)	6
1	0	5	0	1
1	0	4	(0–2)	3
1	0	3	(0–4)	5
1	0	2	(0–6)	7
1	0	1	(0–8)	9
1	0	0	(0–10)	11
0	4	0	0	1
0	3	2	(0–1)	2
0	3	1	(0–3)	4
0	3	0	(0–5)	6
0	2	5	0	1
0	2	4	(0–2)	3
0	2	3	(0–4)	5
0	2	2	(0–6)	7
0	2	1	(0–8)	9
0	2	0	(0–10)	11
0	1	7	(0–1)	2
0	1	6	(0–3)	4
0	1	5	(0–5)	6
0	1	4	(0–7)	8
0	1	3	(0–9)	10
0	1	2	(0–11)	12
0	1	1	(0–13)	14
0	1	0	(0–15)	16

continued

Half-dollars	*Quarters*	*Dimes*	*Nickels*	*Number of Possibilities*
0	0	10	0	1
0	0	9	(0–2)	3
0	0	8	(0–4)	5
0	0	7	(0–6)	7
0	0	6	(0–8)	9
0	0	5	(0–10)	11
0	0	4	(0–12)	13
0	0	3	(0–14)	15
0	0	2	(0–16)	17
0	0	1	(0–18)	19
0	0	0	(0–20)	21

Adding up the numbers in the right-most column gives a total of 292 ways of making change for a dollar using half-dollar coins, quarters, dimes, nickels, and pennies. Counting up all of the possibilities by brute force does work, but it is time consuming and a less-than-elegant way to solve the problem. Additionally, trying to find the number of ways to change higher denominations (say, a $5 bill or $10 bill, not to mention anything higher) by using this method quickly becomes impractical. Clearly, we need an insight in a big way!

Insight

Luckily, there is another way to approach these kinds of problems. Instead of counting out every individual possibility, we can use cleverly chosen polynomials to generate the answer to problems similar to how many ways there are to make change for a dollar.

We will begin by examining the technique of multiplying polynomials. For example,

$$\left(1+x+x^2+x^3\right)\left(1+x^2+x^4+x^6+x^8\right)$$

$$=1+x+2x^2+2x^3+2x^4+2x^5+2x^6+2x^7+2x^8+2x^9+x^{10}+x^{11}.$$

Suppose, though, that we are interested in the terms only up to x^8. Then we could rewrite the product neglecting any terms involving powers higher than the eighth power, as

$$\left(1+x+x^2+x^3\right)\left(1+x^2+x^4+x^6+x^8\right)$$
$$=1+x+2x^2+2x^3+2x^4+2x^5+2x^6+2x^7+2x^8\ldots.$$

Let's apply this idea to the multiplication of

$$(1+x)\left(1+x^2\right)\left(1+x^3\right)\left(1+x^4\right)\left(1+x^5\right)\left(1+x^6\right).$$

If we say that we are interested in the terms in the product only up to x^6, we can rewrite the multiplication, step by step, as

$$(1+x)\left(1+x^2\right)\left(1+x^3\right)\left(1+x^4\right)\left(1+x^5\right)\left(1+x^6\right)$$
$$=(1+x)\left(1+x^2\right)\left(1+x^3\right)\left(1+x^4\right)\left(1+x^5+x^6\ldots\right)$$
$$=(1+x)\left(1+x^2\right)\left(1+x^3\right)\left(1+x^4+x^5+x^6\ldots\right)$$
$$=(1+x)\left(1+x^2\right)\left(1+x^3+x^4+x^5+x^6\ldots\right)$$
$$=(1+x)\left(1+x^2+x^3+x^4+2x^5+2x^6\ldots\right)$$
$$=1+x+x^2+2x^3+2x^4+3x^5+4x^6\ldots.$$

Let's look at the $4x^6$ term in the last multiplication product. The coefficient 4 lets us know that the multiplication process gave an x^6 term four times. These four instances must have come from the following multiplications:

$$x^6,\quad x^5x,\quad x^4x^2,\quad x^3x^2x.$$

Looking at the exponents in each of these terms, we can see that they represent all of the different partitions of the number 6 with *distinct* summands.

If we think back to the dollar bill problem, we may notice that we are essentially asking for the number of partitions of 100 cents, using 50-, 25-, 10-, 5-, and 1-cent pieces. By using polynomials, as we have done above, we would have the number of partitions with *distinct* summands. However, we will not restrict ourselves to using only one of each coin in order to make change for a dollar (because it would be impossible!). So the question becomes whether we can cleverly choose polynomials that will give us the number of partitions of 100 cents without the constraint of the summands being distinct. The answer is yes, given that we use appropriately chosen polynomials.

Let's consider the multiplication of the following:

$$\left(1+x+x^2+x^3+x^4+x^5+x^6+x^7\right)\left(1+x^2+x^4+x^6\right)\left(1+x^3+x^6\right)$$
$$\left(1+x^4\right)\left(1+x^5\right)\left(1+x^6\right)\left(1+x^7\right).$$

For now, let us rewrite the first three terms:

$$1+x+x^2+x^3+x^4+x^5+x^6+x^7$$
$$=1+x^1+x^{1+1}+x^{1+1+1}+x^{1+1+1+1}+x^{1+1+1+1+1}+x^{1+1+1+1+1+1}+x^{1+1+1+1+1+1+1},$$
$$1+x^2+x^4+x^6=1+x^2+x^{2+2}+x^{2+2+2},$$
$$1+x^3+x^6=1+x^3+x^{3+3},$$

and substitute these forms into the multiplication to get

$$\left(1+x^1+x^{1+1}+x^{1+1+1}+x^{1+1+1+1}+x^{1+1+1+1+1}+x^{1+1+1+1+1+1}+x^{1+1+1+1+1+1+1}\right)$$
$$\left(1+x^2+x^{2+2}+x^{2+2+2}\right)\cdot$$
$$\left(1+x^3+x^{3+3}\right)\left(1+x^4\right)\left(1+x^5\right)\left(1+x^6\right)\left(1+x^7\right).$$

Now let's turn our attention to the coefficient of x^7 in the expansion of the above multiplication. We can see that this coefficient can be thought of as the number of ways of choosing one term from each of the polynomials in the multiplication such that the sum of their x coefficients is 7. We could, for example choose x^{1+1+1} from the first polynomial, x^{2+2} from the second polynomial, and 1 from all of the other polynomials. The sum of the coefficients of x for this set of choices is $(1+1+1)+(2+2)+0+0+0+0+0 = 7$.

We have arrived at a sum of 7 by choosing three 1s and two 2s. Thus, we have eliminated the "distinct" summands restriction that we previously faced. Therefore, the coefficient of x^7 in the above multiplication will represent the number of ways of getting a sum of 7 using summands from 1 to 7, where the summands do not have to be distinct. If we wanted the number of ways to get a sum of 7 using, for example, only odd summands, we would simply get rid of the polynomials in the above expression that contribute even summands. Our answer would be the coefficient of x^7 in the expansion of the multiplication

$$\left(1+x^{1}+x^{1+1}+x^{1+1+1}+x^{1+1+1+1}+x^{1+1+1+1+1}+x^{1+1+1+1+1+1}+x^{1+1+1+1+1+1+1}\right)$$
$$\left(1+x^{3}+x^{3+3}\right)\left(1+x^{5}\right)\left(1+x^{7}\right).$$

These examples give rise to the following general principle: Let a, b, c, d, and e be unequal positive integers. Then the coefficient of x^n in the expansion of

$$\left(1+x^{a}+x^{2a}+x^{3a}+\ldots\right)\left(1+x^{b}+x^{2b}+x^{3b}\ldots\right)\left(1+x^{c}+x^{2c}+x^{3c}\ldots\right)$$
$$\left(1+x^{d}+x^{2d}+x^{3d}\ldots\right)\left(1+x^{e}+x^{2e}+x^{3e}\ldots\right).$$

is the number of partitions of n with summands a, b, c, d, and e. Each factor in the multiplication must include all exponents not exceeding n.

Given what we have just discussed, finding the number of ways to change a dollar bill is now a simple task. We simply use the principle discussed above and plug in

$$a=1, \quad b=5, \quad c=10, \quad d=25, \quad e=50.$$

We are looking for the coefficient of x^{100} in the product $P_1P_2P_3P_4P_5$, where

$$P_1=1+x+x^{2}+x^{3}\ldots+x^{99}+x^{100},$$

$$P_2=1+x^{5}+x^{10}+x^{15}\ldots+x^{95}+x^{100},$$

$$P_3=1+x^{10}+x^{20}+x^{30}\ldots+x^{90}+x^{100},$$

$$P_4=1+x^{25}+x^{50}+x^{75}+x^{100},$$

$$P_5=1+x^{50}+x^{100}.$$

We need to calculate the products only up to the 100th power. We are not concerned with anything larger than that. So, let's begin:

$$P_4P_5 = 1 + x^{25} + 2x^{50} + 2x^{75} + 3x^{100} \ldots,$$

$$P_3P_4P_5 = 1 + x^{10} + x^{20} + x^{25} + x^{30} + x^{35} + x^{40} + x^{45} + 3x^{50} + x^{55} + 3x^{60} + x^{65}$$

$$+3x^{70} + 3x^{75} + 3x^{80} + 3x^{85} + 3x^{90} + 3x^{95} + 6x^{100} \ldots,$$

$$P_2P_3P_4P_5 = 1 + x^5 + 2x^{10} + 2x^{15} + 3x^{20} + 4x^{25} + 5x^{30} + 6x^{35} + 7x^{40} + 8x^{45}$$

$$+11x^{50} + 12x^{55} + 15x^{60} + 16x^{65} + 19x^{70} + 22x^{75} + 25x^{80}$$

$$+28x^{85} + 31x^{90} + 34x^{95} + 40x^{100} \ldots.$$

Multiplying through by P_1 does not need to be done rigorously because we are concerned only with the coefficient of x^{100}. Each term in the product $P_2P_3P_4P_5$ will contribute once to the coefficient of x^{100} in $P_1P_2P_3P_4P_5$. So, this coefficient can be calculated by adding all of the coefficients in $P_2P_3P_4P_5$. This gives

1 + 1 + 2 + 2 + 3 + 4 + 5 + 6 + 7 + 8 + 11 + 12 + 15 + 16 + 19 + 22 + 25 + 28 + 31 + 34 + 40 = 292.

Thus, there are 292 ways to change a dollar bill using half dollar coins, quarters, dimes, nickels, and pennies. This exactly matches (as it must) the answer we got by using the brute force method.

We now can see the power of approaching this problem by using polynomials. We could just as easily have figured out the number of ways to change a \$5 bill or a \$10 bill or anything still higher than that. Alternatively, we could have easily determined how many ways there are to change a dollar without using quarters (by simply restricting the summands to 50, 10, 5, and 1). The generality of the approach of generating polynomials to solve these problems makes it far superior to a simple enumeration of possibilities.

Tools Used and Developed

- Using multiplication of polynomials to find the number of ways summands can be combined to give a certain total.

3

Number Theory

Problem 3.1

Evaluate the infinite continued fraction

$$1+\cfrac{1}{1+\cfrac{1}{1+\cfrac{1}{1+\cfrac{1}{1+\dots}}}}.$$

Although I do not know of any brute force method for solving this problem, the solution I do know requires the following clever insight. Let x equal the fraction above:

$$x=1+\cfrac{1}{1+\cfrac{1}{1+\cfrac{1}{1+\cfrac{1}{1+\dots}}}}.$$

Because the fraction has the same repeating form and is infinite in extent, we can replace the entire denominator in the second term by x, and write

$$x=1+\frac{1}{x}.$$

We can now multiply through by x to clear the fraction, and we get

$$x^2 = x + 1, \text{ or } x^2 - x - 1 = 0.$$

This last equation can be solved by the quadratic formula to give the two solutions,

$$x = \frac{1 \pm \sqrt{5}}{2}.$$

Since the continued fraction is obviously positive, we can say that $x = (1+\sqrt{5})/2$, and we discard $x = (1-\sqrt{5})/2$ because it is negative. Therefore, we have the odd result that

$$1+\cfrac{1}{1+\cfrac{1}{1+\cfrac{1}{1+\cfrac{1}{1+\ddots}}}} = \frac{1+\sqrt{5}}{2}.$$

This number, $(1+\sqrt{5})/2$, turns out to be very important in math. It is known as the golden ratio. It is traditionally referred to as Φ (the Greek letter *phi*). This number turns up in many places, ranging from biology to art. Our interest, however, is its relationship to the famous Fibonacci sequence—perhaps the most famous example of a recurrence relation, or recursive sequence, meaning that each term in the sequence depends in some way on the terms before it. A recursive sequence is defined by two properties: the rule that describes how to get the next term from the terms that have come before, and the "seed" values that start the sequence going. The Fibonacci sequence, for example, is defined as

$$F_n = F_{n-1} + F_{n-2}.$$

This means that the current term is the sum of the two previous terms—that's the rule. To start the Fibonacci sequence, we need the seed values, which are $F_1 = 1$ and $F_2 = 1$. From the rule for getting the next term, we know that $F_3 = F_2 + F_1$; or $F_3 = 1 + 1 = 2$. Now that we have F_3, we can use it and F_2 to generate F_4: $F_4 = F_3 + F_2$, or $F_4 = 2 + 1 = 3$.

In this way, we can keep generating terms of the Fibonacci sequence as follows:

F_1	F_2	F_3	F_4	F_5	F_6	F_7	F_8	F_9	F_{10}	F_{11}	F_{12}
1	1	2	3	5	8	13	21	34	55	89	144

Let's get back to what this has to do with Φ. Let us consider the ratio of successive terms of the Fibonacci sequence, and see how this changes with n. For a geometric sequence,

$$a \quad ar \quad ar^2 \quad ar^3 \quad ar^4 \cdots,$$

the ratio of the successive terms is fixed: r. The Fibonacci sequence is not quite a geometric sequence. The ratio of the terms changes as we progress along the sequence. For example,

$$\frac{F_2}{F_1} = \frac{1}{1}, \quad \frac{F_3}{F_2} = \frac{2}{1}, \quad \frac{F_4}{F_3} = \frac{3}{2}, \quad \frac{F_5}{F_4} = \frac{5}{3}.$$

Let us look first at $F_3/F_2 = 2/1$. We can rewrite this as

$$\frac{F_3}{F_2} = 2 = 1 + \frac{1}{1}. \tag{3.1}$$

We can also write $F_4/F_3 = 3/2$ as

$$\frac{F_4}{F_3} = \frac{2+1}{2} = 1 + \frac{1}{2}. \tag{3.2}$$

However, since 2 can be written as 1 + 1/1 (see Equation (3.1)), if we substitute this into the denominator of 1/2 in Equation (3.2), we get

$$\frac{F_4}{F_3} = \frac{3}{2} = 1 + \frac{1}{2} = 1 + \cfrac{1}{1 + \cfrac{1}{1}}. \tag{3.3}$$

Let's look at $F_5/F_4 = 5/3$, which can be written as $F_5/F_4 = (3 + 2)/3 = 1 + 2/3$. However, 2/3 is just the reciprocal of 3/2 (see Equation (3.3)), so

$$\frac{F_5}{F_4} = \frac{5}{3} = 1 + \frac{2}{3} = 1 + \cfrac{1}{1 + \cfrac{1}{1 + \cfrac{1}{1}}}.$$

We can now see the pattern:

$$\frac{F_n}{F_{n-1}} = \frac{F_{n-1} + F_{n-2}}{F_{n-1}} = 1 + \frac{F_{n-2}}{F_{n-1}} = 1 + \cfrac{1}{\cfrac{F_{n-1}}{F_{n-2}}}.$$

But F_{n-1}/F_{n-2} is just a fraction of the form

$$1 + \cfrac{1}{1 + \cfrac{1}{1 + \cfrac{1}{1 + \cfrac{1}{1 + \dots}}}}.$$

Thus, we see that Φ is the limit of the ratio of the successive terms of the Fibonacci sequence, or F_n/F_{n-1}, as $n \to \infty$. Remembering that Φ is a solution to the equation $x^2 - x - 1 = 0$, we see how this equation is intimately related to the Fibonacci sequence. In fact, we can call it the characteristic equation of the Fibonnaci sequence. We will come back to this idea later.

Using the ideas that we just discussed, let's try to tackle the following problem using similar insights. Evaluate the following infinite expression:

$$\sqrt{1 + \sqrt{1 + \sqrt{1 + \sqrt{1 + \dots}}}}.$$

To solve this continued radical, we use the exact same insight as before; let

$$x = \sqrt{1 + \sqrt{1 + \sqrt{1 + \sqrt{1 + \dots}}}}.$$

Because the radical continues forever, we can rewrite the equation now as

$$x = \sqrt{1 + x}.$$

Squaring both sides, we get $x^2 = x + 1$, or $x^2 - x - 1 = 0$, the same equation we just solved, and because the radical is positive, we can say that

$$x = \frac{1+\sqrt{5}}{2}.$$

This is the same as the value of our continued fraction, so we have proven the curious result that

$$1+\cfrac{1}{1+\cfrac{1}{1+\cfrac{1}{1+\cfrac{1}{1+\ddots}}}} = \sqrt{1+\sqrt{1+\sqrt{1+\sqrt{1+\dots}}}}.$$

Tools Used and Developed

- Substituting for x to solve for x; that is, solving for x in terms of itself.
- Definition of and familiarity with the Fibonacci sequence.

Problem 3.2

Every day at school, Jo climbs a flight of six stairs. Jo can take the stairs one, two, or three at a time. For example, Jo can climb three steps, then one step, then two steps. In how many ways can Jo climb the stairs?

Trying to count the number of ways by just listing them arbitrarily becomes very confusing. Therefore, if we are going to count, we need an organized way to do it. For six stairs, one approach is to think about the ways we can climb the stairs in terms of combinations of different step values. For example, if we are going to take only single steps, then there is only one way to climb the stairs: 1-1-1-1-1-1.

Now let's think about combinations of one and two steps at a time. Here, we start with all 1-steps except for one 2-step. Thus, we would have some permutation of 1-1-1-1-2. How many of these are there? This may not be so easy to determine, but here is an insight on how to handle a problem like this. Think of this sequence: 010101010. The 1s represent the four single steps we will take, and the 0 is a placeholder where we can insert the 2-step that we will take. We see that it can come before, after, or anywhere in between the 1-steps, so there are five possibilities for this. An easier way to see this is that 1-1-1-1-2 has five number spots, and the 2 can be in any of those five. For example, let's think about two 1s and two 2s, such as 1-2-1-2. Here, we can think of four slots and we have to pick two of them for the 2s, so this is $\binom{4}{2}$, or 6. Explicitly, these are 1-1-2-2, 1-2-1-2, 1-2-2-1, 2-1-1-2, 2-1-2-1, and 2-2-1-1. This gives us a total of 11 possibilities using some combination of 1s and 2s. There are no other ways to go with just 1s and 2s.

If we now consider just 2s, that's easy—just 2-2-2. For 1 and 3, we must have some permutation of three 1s and one 3, or some permutation of 1-1-1-3. By the reasoning above, there are just four possibilities. Considering just 2s and 3s is easy. There is no way to climb six stairs using a combination of 2 and 3. For 1, 2, and 3, we have some permutation of 1-2-3, since each of 1, 2, and 3 has to occur only once, so this is also easy; it is just 3!, or 6. And just 3-steps can be done in only one way: 3-3.

Thus, we can make a table of the number of combinations, which shows a total of 24 ways.

Combination of Step Types Used	*Number of Combinations*
1	1
1, 2	11
1, 3	4
2	1
2, 3	0
1, 2, 3	6
3	1

The official solution also uses a method of systematic counting that is almost the same. It focuses on the number of stair moves to get the six stairs. Then, the categories break up a bit differently. For example, to get the six stairs done in four moves, one possibility would be 1-2-1-2, and another would be 1-1-1-3. All of the permutations of these, of course, would also count. This way, we would get the following table. Once again, this gives 24 ways to climb the six steps.

Number of Stair Moves	*Stair Move Sequence*	*Count*
6	1-1-1-1-1-1	1
5	1-1-1-1-2 and its permutations (5)	5
4	1-1-1-3 and its permutations (4)	
	1-1-2-2 and its permutations (6)	10
3	1-2-3 and its permutations (6)	
	2-2-2	7
2	3-3	1

We can see that if the number of steps n grows even a bit larger, say $n = 10$, the systematic counting method becomes very cumbersome and impractical. If we want to be able to solve this problem for large values of n, we need an insight.

Insight 1

To develop a more insightful approach, let us use one of the greatest strategies in problem solving: solve an easier problem! This can get us on our way. Let us begin by looking at the number of ways to climb n stairs when we can take only one step or two steps at a time (not one, two, or three, as we discussed above). The easiest thing is to work backwards (another excellent problem solving strategy). Assume we are at step n. There are only two paths we could have taken to get here: either we were at step $n - 1$ and took a single step, or we were at step $n - 2$, and took a double step; no other possibilities exist. Therefore, the number of ways to get to step n, call that S_n, is the sum of the possible ways to get to step $n - 1$ and the ways to get to step $n - 2$. As an equation:

$$S_n = S_{n-1} + S_{n-2}.$$

This should look very familiar to us. It is a recurrence sequence, and one that has the same form as the famous Fibonacci sequence we encountered previously, $F_n = F_{n-1} + F_{n-2}$. Traditionally, the Fibonacci sequence is defined starting at $F_1 = 1$ and $F_2 = 1$, and everything else just follows: $F_3 = 2$, $F_4 = 3$, $F_5 = 5$, and so on.

Here, we start at step 0, so we will define $S_0 = 1$. From step 0, there is only one way to get to step 1: just take one step. Thus, $S_1 = 1$. Starting at step 0, there are two ways to get to step 2: two single steps (1-1), or a double step (2).

To get to step 3, we either have to be at step 2 and take one step, or at step 1 and take a double step, so $S_3 = S_2 + S_1$; or $S_3 = 3$. And so on, with $S_n = S_{n-1} + S_{n-2}$. Thus, we have $S_0 = 1$, $S_1 = 1$, $S_2 = 2$, $S_3 = 3$, ...

We see, then, that this is the same set of numbers as the Fibonacci sequence, except that it is "one behind." In other words, $S_n = F_{n+1}$. Thus, to see how many ways we can climb n stairs, we need to calculate the $(n + 1)$st Fibonacci number.

Following the same logic, when we are allowed to take the stairs using one, two, or three steps, as in our original problem, we can see that to get to step n, we could be at step $n - 1$ and take a single step, at step $n - 2$ and take a double step, or at step $n - 3$ and take a triple step. Thus, for our original problem,

$$S_n = S_{n-1} + S_{n-2} + S_{n-3}.$$

This is another recurrence relationship, very similar to the Fibonacci relationship, except that each term is made up of the sum of the previous *three* terms. Now that we have the recurrence relationship, all we need are the beginning seed values to start the machine.

We already know that $S_1 = 1$, and $S_2 = 2$. There are four ways, however, to get to step 3: (1-1-1; 1-2; 2-1; 3). Thus, $S_3 = 4$. Notice that this happens to equal $S_0 + S_1 + S_2$, just as our recurrence relationship suggests. We can now quickly build up to any number of steps:

$$S_4 = S_3 + S_2 + S_1 = 4 + 2 + 1 = 7.$$

$$S_5 = S_4 + S_3 + S_2 = 7 + 4 + 2 = 13.$$

$$S_6 = S_5 + S_4 + S_3 = 13 + 7 + 4 = 24.$$

Thus, there are 24 ways to climb six steps, as we had found previously. Also, we can quickly build up to the number of ways to climb 10 steps, for instance. Following the above algorithm, $S_7 = 44$, $S_8 = 81$, $S_9 = 149$, and $S_{10} = 274$.

Insight 2: Even More Elegant

To solve our easier problem in a one-step fashion, we can use the characteristic polynomial of the Fibonnaci sequence, $x^2 - x - 1$, to derive a closed-form equation for the nth Fibonacci number, F_n. That way, we don't need to calculate all of the Fibonacci numbers just to get to F_n. When we solve the characteristic polynomial $x^2 - x - 1 = 0$ by using the quadratic equation, we get two solutions, $(1 \pm \sqrt{5})/2$. Let us call these two solutions α and τ (as is traditional). Thus, we let $\alpha = (1+\sqrt{5})/2$, and we let $\tau = (1-\sqrt{5})/2$. Since α is a solution to $x^2 - x - 1 = 0$, we can write $\alpha^2 - \alpha - 1 = 0$, or $\alpha^2 = \alpha + 1$.

Let's now find an expression for α^3. We can rewrite this as $\alpha^2(\alpha)$. Since $\alpha^2 = \alpha + 1$, we can write $\alpha^3 = (\alpha + 1)(\alpha) = \alpha^2 + \alpha = (\alpha + 1) + \alpha = 2\alpha + 1$. Let's follow a similar procedure for α^4: $\alpha^4 = \alpha^3(\alpha)$, but we now know that $\alpha^3 = 2\alpha + 1$, so we have $\alpha^4 = (2\alpha + 1)(\alpha) = 2\alpha^2 + \alpha = 2(\alpha + 1) + \alpha = 3\alpha + 2$. A pattern is beginning to emerge. Let us consolidate our impression by going through the process one more time for α^5: $\alpha^5 = \alpha^4(\alpha)$, but we now know that $\alpha^4 = 3\alpha + 2$, so we have $\alpha^5 = (3\alpha + 2)(\alpha) = 3\alpha^2 + 2\alpha = 3(\alpha + 1) + 2\alpha = 5\alpha + 3$.

We can now see the pattern:

$$\alpha^n = F_n\alpha + F_{n-1}.$$

This, of course, will require proof, and it is ideal for mathematical induction (we discussed this technique in Problem 1.3).

We have already established the case for some n (for $n = 3$, 4, and 5). Now we need to prove that if it holds for n, then it holds for $n + 1$. So, let us calculate α^{n+1}.

$$\begin{aligned}
\alpha^{n+1} &= \alpha^n(\alpha) \\
&= (F_n\alpha + F_{n-1})(\alpha) \\
&= F_n\alpha^2 + F_{n-1}\alpha \\
&= F_n(\alpha + 1) + F_{n-1}\alpha \\
&= (F_n + F_{n-1})\alpha + F_n.
\end{aligned}$$

Since $F_{n+1} = F_n + F_{n-1}$, we can write $\alpha^{n+1} = F_{n+1}\alpha + F_n$; this completes our inductive proof, and we can state with confidence that

$$\alpha^n = F_n\alpha + F_{n-1}. \tag{3.4}$$

We know that τ is also a root of $x^2 - x - 1 = 0$, so the same argument also proves that

$$\tau^n = F_n\tau + F_{n-1}. \tag{3.5}$$

If we now subtract Equation (3.5) from Equation (3.4), we get

$$\alpha^n - \tau^n = F_n(\alpha - \tau). \tag{3.6}$$

Notice how the F_{n-1} terms canceled out in the subtraction.

Thus, from Equation (3.6), we now have a formula for F_n:

$$F_n = \frac{\alpha^n - \tau^n}{\alpha - \tau}, \tag{3.7}$$

where $\alpha = (1+\sqrt{5})/2$ and $\tau = (1-\sqrt{5})/2$. Plugging α and τ into the denominator, we can further simplify Equation (3.7) to the following remarkable formula to directly calculate the nth Fibonacci number:

$$F_n = \frac{\alpha^n - \tau^n}{\sqrt{5}}. \tag{3.8}$$

Remarkably, this formula, called the Binet formula, actually has an integer value for any value of n; moreover, that value, as we have just derived, is F_n. Thus, we have a one-step solution for our easier stairs problem, where the student can take the stairs one or two at a time.

We have already established that the solution for our harder problem, where the student can take the stairs 1, 2, or 3 at a time, is obtained by calculating the corresponding Tribonacci number:

$$T_n = T_{n-1} + T_{n-2} + T_{n-3}.$$

It turns out that there is also an analogous formula for calculating the nth Tribonacci number, based on the solutions of its characteristic polynomial, $x^3 - x^2 - x - 1 = 0$. This derivation is too complex for us to do, but we can give the final result:

$$\begin{aligned} T_n &= \frac{\alpha^{n+1}}{(\alpha-\beta)(\alpha-\gamma)} + \frac{\beta^{n+1}}{(\beta-\alpha)(\beta-\gamma)} + \frac{\gamma^{n+1}}{(\gamma-\alpha)(\gamma-\beta)} \\ &= \frac{\alpha^n}{-\alpha^2+4\alpha-1} + \frac{\beta^n}{-\beta^2+4\beta-1} + \frac{\gamma^n}{-\gamma^2+4\gamma-1}, \end{aligned}$$

where (α,β,γ) are the three roots of the polynomial $P(x) = x^3 - x^2 - x - 1$.

Tools Used and Developed

- The definition of Fibonacci numbers, developed in Problem 3.1.
- The Binet formula for the calculation of Fibonacci numbers.
- Generalization of the Fibonnacci numbers to such things as the Tribonacci numbers.
- Working backward.

Problem 3.3

Let $a_0 = a_1 = 1$ and $a_{n+1} = 7a_n - a_{n-1} - 2$. Calculate a_5.

These recursive definitions are painful to churn through, but certainly fairly easy to understand. We just grind through it.

$$a_2 = 7a_1 - a_0 - 2 = 7(1) - 1(1) - 2 = 4,$$

$$a_3 = 7a_2 - a_1 - 2 = 7(4) - 1(1) - 2 = 25,$$

$$a_4 = 7a_3 - a_2 - 2 = 7(25) - 1(4) - 2 = 169,$$

$$a_5 = 7a_4 - a_3 - 2 = 7(169) - 1(25) - 2 = 1{,}156.$$

Since this took us only a few minutes, we now try to calculate a_{10}. But we don't want to spend a whole lot of time doing recursive sequence calculations, so we need an insight.

Insight

We really have no good way to begin, so we look at our answers so far, and something a bit peculiar catches our attention:

$$a_2 = 4 = 2^2;\ a_3 = 25 = 5^2;\ a_4 = 169 = 13^2.$$

That's a little unusual. The first three terms we calculated are perfect squares. We don't immediately recognize 1,156, so we check it as well and we find that $a_5 = 1{,}156 = 34^2$.

This is interesting. As we look at these base numbers, we find them to be vaguely familiar. Thinking a bit harder, we notice that they seem to be the squares of every other Fibonacci number, since the Fibonacci sequence is

$$1, \quad 1, \quad 2, \quad 3, \quad 5, \quad 8, \quad 13, \quad 21, \quad 34, \quad \ldots,$$

where for the counting in our problem we will define it by the recursion

$$F_1 = 1; \quad F_2 = 1; \quad F_{n+1} = F_n + F_{n-1}.$$

If this pattern continues, we can conjecture that $a_n = F_{2n-1}^2$. How would we prove this conjecture? Since we already know, given our labeling of the Fibonacci numbers, that $a_2 = 4 = F_3^2$, $a_3 = 25 = F_5^2$, $a_4 = 169 = F_7^2$, $a_5 = 1156 = F_9^2$, it would seem natural to try an inductive proof, where we use the fact that for several initial cases, we know that $a_k = F_{2k-1}^2$.

Let's start with the recursive definition of our sequence and write the following equalities for a_{n+1} and for a_n directly from the definition:

$$a_{n+1} = 7a_n - a_{n-1} - 2, \tag{3.9}$$

$$a_n = 7a_{n-1} - a_{n-2} - 2. \tag{3.10}$$

Subtracting Equation (3.10) from Equation (3.9), we get,

$$a_{n+1} - a_n = 7a_n - 8a_{n-1} + a_{n-2}, \text{ or } a_{n+1} = 8a_n - 8a_{n-1} + a_{n-2}. \tag{3.11}$$

Using our induction hypothesis, $a_k = F_{2k-1}^2$, we can rewrite Equation (3.11) as

$$a_{n+1} = 8F_{2n-1}^2 - 8F_{2n-3}^2 + F_{2n-5}^2. \tag{3.12}$$

To complete our proof, we would then need to show that

$$8F_{2n-1}^2 - 8F_{2n-3}^2 + F_{2n-5}^2 = F_{2n+1}^2.$$

Let us start with the definition of the Fibonacci sequence and write some of the terms as the sum of lower-order terms:

$$F_{m+2} = F_{m+1} + F_m,$$

$$F_{m+1} = F_m + F_{m-1},$$

$$F_m = F_{m-1} + F_{m-2}.$$

By using these identities, we can write that

$$F_{m+2} = F_m + F_{m-1} + F_m = 2F_m + F_{m-1}. \tag{3.13}$$

Also, since $F_{m-1} = F_m - F_{m-2}$, we can substitute this for F_{m-1} in Equation (3.13), to get

$$F_{m+2} = 3F_m - F_{m-2}, \tag{3.14}$$

and reciprocally,

$$F_{m-2} = 3F_m - F_{m+2}. \tag{3.15}$$

The purpose of these manipulations is to write the Fibonacci identities in terms of every other Fibonacci number, since the pattern we have noticed involves the squares of the odd Fibonacci numbers.

Now let's go back to Equation (3.12). We would like to substitute for F_{2n-5} by using the identity in Equation (3.15), $F_{m-2} = 3F_m - F_{m+2}$. We equate the indices by letting $m - 2 = 2n - 5$. Thus, $m = 2n - 3$ and $m + 2 = 2n - 1$, and we can now write

$$F_{2n-5} = 3F_{2n-3} - F_{2n-1}. \tag{3.16}$$

Substituting Equation (3.16) into Equation (3.12), $a_{n+1} = 8F_{2n-1}^2 - 8F_{2n-3}^2 + F_{2n-5}^2$, we have $a_{n+1} = 8F_{2n-1}^2 - 8F_{2n-3}^2 + (3F_{2n-3} - F_{2n-1})^2$. Expanding the binomial, we have

$$\begin{aligned} a_{n+1} &= 8F_{2n-1}^2 - 8F_{2n-3}^2 + 9F_{2n-3}^2 + F_{2n-1}^2 - 2(3)F_{2n-3}F_{2n-1} \\ &= 9F_{2n-1}^2 + F_{2n-3}^2 - 2(3)F_{2n-3}F_{2n-1} \\ &= (3F_{2n-1} - F_{2n-3})^2. \end{aligned}$$

However, going back to Equation (3.14), $F_{m+2} = 3F_m - F_{m-2}$, we see that with $m = 2n - 1$, $3F_{2n-1} - F_{2n-3} = F_{2n+1}$, so we have proven that $a_{n+1} = (3F_{2n-1} - F_{2n-3})^2 = F_{2n+1}^2$, or that, in general, $a_n = F_{2n-1}^2$ for our given recursive sequence. Thus, for the problem of calculating a_{10}, we know that this will equal F_{19}^2.

In Problem 3.2, we learned how we can calculate the Fibonacci numbers directly. If we let $\alpha = (1+\sqrt{5})/2$ and $\tau = (1-\sqrt{5})/2$, then amazingly, $F_n = (\alpha^n - \tau^n)/(\alpha - \tau)$, which is Equation (3.7), the Binet formula, $F_n = (\alpha^n - \tau^n)/\sqrt{5}$, which is Equation (3.8). Thus, plugging into Equation (3.7), we have $F_{19} = 4{,}181$, and $a_{10} = (4{,}181)^2 = 17{,}480{,}761$.

The insight used in this problem is what we are calling a type 3 insight, where, by trial and error, or case work, we happen to notice a pattern, and then prove that this pattern holds. For me, this is the least fun type of insight, but is often incredibly useful, as we saw here.

Tools Used and Developed

- Our knowledge of the Fibonacci numbers and manipulating their definition, $F_{n+1} = F_n + F_{n-1}$, to give us new Fibonacci identities.

Problem 3.4

An interesting question is to figure out how many ways an integer can be represented as the sum of one or more integers, where order matters. In how many ways can 10 be represented?

Let's look at some examples. The number 3 can be written as: 3, 1 + 2, 2 + 1, and 1 + 1 + 1. Order matters in this problem, so 1 + 2 is considered different from 2 + 1, and 3 can be written in four ways. In how many ways can 4 be represented? We can list them, being careful not to lose track: 4, 1 + 3, 3 + 1, 1 + 1 + 2, 1 + 2 + 1, 2 + 1 + 1, 1 + 1 + 1 + 1, 2 + 2. So 4 can be represented in eight different ways.

Back to the original problem, in how many ways can 10 be represented? This will require a monumental amount of very tedious and meticulous record-keeping to make sure we accurately tally all of the combinations. Furthermore, we wouldn't have an easy way to know whether we've missed one along the way. We need to try to find an analytical solution.

We might start by looking at the representations of 3 and 4, which we have listed explicitly above, hoping to find a way to move from one to the other. Since 4 is 1 more than 3, we can think of taking each of the representations of 3, and adding 1. However, since order matters, we would have to add 1 to both the front and the back of each sequence for 3, as shown in the following table.

Representations of 3	*Representations of 4*
3	1 + 3, 3 + 1
1 + 2	1 + 1 + 2, 1 + 2 + 1
2 + 1	1 + 2 + 1, 2 + 1 + 1
1 + 1 + 1	1 + 1 + 1 + 1

This strategy, although it's a nice way to think, doesn't work. First, some of the permutations are duplicated (as underlined in the table). Second, we miss certain key representations, such as 2 + 2, and 4 itself. Clearly, we need an insight.

Insight

The first insight, really a "mini-insight," is that we are not being asked to list the representations, but to just to figure out how many there are. Therefore, we need to "keep our eye on the ball." Perhaps we can count the representations without explicitly listing them.

Our larger insight is that we realize that if we break up the integer into a sum of smaller integers, those smaller integers can also be broken down until, ultimately, each integer is the sum of a certain number of 1s. Thus, ultimately, an integer n is the sum of n 1s, and all representations can be thought of as different ways of grouping these 1s.

Let's look at this idea with $n = 4$:

1 _ 1 _ 1 _ 1

We have written the four 1s with three spaces between them. We can now fill each space with a plus (+) or a slash (/). If there is a plus, we add the 1s together to make a bigger number, and if there is a slash, we keep them separate. This will be a little confusing, because above, we used the + sign in listing each representation as the sum of separate integers. In addition, the slash is just a spacer, not a sign for division. However, a couple of examples should take away any confusion.

If we look at 1 / 1 / 1 / 1, this means to take four separate 1s. In other words, this means to represent 4 as 1 + 1 + 1 + 1. If we take 1 / 1 + 1 / 1, this means to take 1, then add 1 + 1 to get 2, then take another 1, to get 1 + 2 + 1 as a representation of 4. Therefore, we set up a one-to-one correspondence between how we want to distribute pluses and slashes with each of the representations of 4 we had listed previously, as shown in the following table.

Representation	*Stands for*
1 / 1 / 1 / 1	1 + 1 + 1 + 1
1 + 1 + 1 / 1	3 + 1
1 + 1 / 1 + 1	2 + 2
1 / 1 + 1 / 1	1 + 2 + 1
1 + 1 / 1 / 1	2 + 1 + 1
1 / 1 / 1 + 1	1 + 1 + 2
1 / 1 + 1 + 1	1 + 3
1 + 1 + 1 + 1	4

As we can see, each distribution of + and / corresponds to exactly one way of representing 4. Since we have three spaces between the four 1s that ultimately make up 4, and in each space, we independently have two choices of what to insert (+ or /), this gives us $2 \times 2 \times 2 = 8$ possible ways of representing 4 according to our problem conditions.

This can be easily generalized for any integer n. Following the reasoning above, we would list n 1s, separated by $(n - 1)$ spaces. In each of these spaces, we could put a + or a /. Each combination of +s and /s filling these $(n - 1)$ spaces corresponds to a unique representation of n. Since there are $(n - 1)$ spaces and two choices for each space, this means that there are 2^{n-1} ways of representing n as the sum of one or more integers, where order matters.

This fits with what we found with our explicit listings for the small cases of $n = 3$ ($2^{3-1} = 2^2 = 4$ ways) and $n = 4$ ($2^{4-1} = 2^3 = 8$ ways). Thus, for $n = 10$, we would have $2^{10-1} = 2^9 = 512$ ways.

It's lucky that we didn't try to explicitly list those out.

Tools Used and Developed

- The multiplication principle in combinatorics: when we have two choices for each of n events, and the choice for each event is independent of the others, we have 2^n possible outcomes.
- The combinatorial tool of using spaces between numbers as an aid to counting how to group them.

Problem 3.5

A hallway in a school has 10 lockers. All the lockers are open on Monday morning. The first student comes in and closes all the lockers. The second student comes in and opens every other locker (lockers 2, 4, 6, 8, and 10). The third student comes in and changes the closed/open positions of all the lockers divisible by 3 (lockers 3, 6, and 9; i.e., if they are open, he closes them, and if closed, he opens them). The fourth student does the same with lockers divisible by 4, changing their status. This continues for all 10 students. After the tenth student has come through, what lockers are closed?

The table below (with C = closed and O = open) helps us to visualize the states of the lockers after each student does the prescribed locker maneuver.

	Locker Number									
	1	*2*	*3*	*4*	*5*	*6*	*7*	*8*	*9*	*10*
Initial status	O	O	O	O	O	O	O	O	O	O
Student 1	C	C	C	C	C	C	C	C	C	C
Student 2	C	O	C	O	C	O	C	O	C	O
Student 3	C	O	O	O	C	C	C	O	O	O
Student 4	C	O	O	C	C	C	C	C	O	O
Student 5	C	O	O	C	O	C	C	C	O	C
Student 6	C	O	O	C	O	O	C	C	O	C
Student 7	C	O	O	C	O	O	O	C	O	C
Student 8	C	O	O	C	O	O	O	O	O	C
Student 9	C	O	O	C	O	O	O	O	C	C
Student 10	C	O	O	C	O	O	O	O	C	O

Let's make sure we understand what has happened here. As an example, student 5 will change only the positions of those lockers divisible by 5, or lockers 5 and 10. Thus, student 5 leaves lockers 1–4 and lockers 6–9 in whatever position they were previously, but changes locker 5 from closed to open and locker 10 from open to closed.

Looking at the above table, we see that after all ten students have passed through the hallway, only three lockers remain closed: lockers 1, 4, and 9. With only ten lockers, our table helps us easily find the answer. But what if there were 100 lockers? Is there an insight that will help us quickly deduce the answer?

Insight

We might notice that locker numbers 1, 4, and 9 are the only ones here that are perfect squares. Could that have something to do with it? Perhaps, but how can we show that?

Since each student is changing the position of only those lockers that are divisible by their own student number, let's take a look at the lockers and their divisors to see if we recognize a pattern in the table below.

Locker Number	*Divisors of Locker Number*	*Final Position of Locker*
1	**1**	Closed
2	1,2	Open
3	1,3	Open
4	**1,2,4**	Closed
5	1,5	Open
6	1,2,3,6	Open
7	1,7	Open
8	1,2,4,8	Open
9	**1,3,9**	Closed
10	1,2,5,10	Open

We notice something in this table about the parity of the number of times the lockers changed status (i.e., whether it was an even or odd number of times). Lockers 1, 4, and 9 (the only lockers that remained closed after the last student had passed through the hallway) are the only ones with an odd number of divisors; the remaining lockers, which ended up in an open position, all have an even number of divisors.

What does that tell us? An even number of divisors means that these lockers were touched an even number of times, so the initial state and the final states of those lockers will be the same. An odd number of divisors means that the locker's final state will be different than its initial state. Since we know that factors always come in pairs, except in the case of perfect squares (where the square root of the perfect square would be listed only once as a factor), we can deduce that locker numbers that are perfect squares will be the only ones whose end positions differ from their starting positions.

So if we had to figure this out for 100 lockers, we don't need to make an extensive table. We only need to know the perfect squares between 1 and 100 (1, 4, 9, 16, 25, 36, 49, 64, 81, 100), and only 10 lockers would be closed—those with the corresponding numbers.

Tools Used and Developed

- Recognizing the effects of parity.

Problem 3.6

To the ancient Greeks, mathematics was closely linked to both philosophy and mysticism. Numbers were thought to possess special properties. One of the most special of these properties was for a number to be "perfect," which means the number is equal to the sum of its *proper* divisors. For example, 6 = 1 + 2 + 3, so 6 is a *perfect* number. Proper divisors include all the divisors, starting at 1, but do not include the number itself (e.g., 6 is a divisor of 6, but is not considered a *proper* divisor). Find the next perfect number after 6.

The procedure is quite straightforward. We just go through the numbers, and make a list of their proper divisors, and check whether they sum to the number. Starting with 6, and moving forward, we get the following table.

Number	*Proper Divisors*	*Sum*
6	1, 2, 3	6
7	1	1
8	1, 2, 4	7
9	1, 3	4
10	1, 2, 5	8
11	1	1
12	1, 2, 3, 4, 6	16
13	1	1
14	1, 2, 7	10
15	1, 3, 5	9
16	1, 2, 4, 8	15
17	1	1
18	1, 2, 3, 6, 9	21
19	1	1
20	1, 2, 4, 5, 10	22
21	1, 3, 7	11
22	1, 2, 11	14
23	1	1
24	1, 2, 3, 4, 6, 8, 12	36
25	1, 5	6
26	1, 2, 13	16
27	1, 3, 9	13
28	1, 2, 4, 7, 14	28

Finally, we see (if we have not made an addition mistake somewhere) that the next perfect number after 6 is 28. What is another perfect number after 28?

Finding such a number is problematic. First, we don't know whether there is another perfect number. Second, even though the method is easy, it is rather tedious and time consuming to list all the divisors and then sum them all up. We need an insight.

Insight 1

We need a faster way to reach the sum of the proper divisors of a number N so we can check whether it's the same as N. The first insight is to realize that we can find a quicker way to sum all of the divisors (not just the proper ones). This means all of the proper divisors as well as N. Therefore, if a number is perfect, then the sum of all the divisors is $2N$, because all of the proper divisors sum to N and then we add N as the last divisor.

How can we sum all of the divisors other than by listing them out? We know that every number can be factorized into a product of 1 and some primes raised to various powers—this is called prime factorization. Let us say that a number N can be prime factorized into $p^s q^t$, where p and q are prime numbers. Then we know that $1, p, p^2, p^3, \ldots, p^s$ are all divisors of N. We also know that $1, q, q^2, q^3, \ldots, q^t$ are also all divisors of N. Any other divisor of N can be only some product of one number from each list, such as pq^2, p^3q^2, and so on. Thus, the sum of all the divisors can be written as

$$\left(1+p+p^2+p^3+\ldots+p^s\right)\left(1+q+q^2+q^3+\ldots+q^t\right).$$

If we multiply out the parentheses, we see that we would get each individual term in the lists $1, p, p^2, p^3, \ldots, p^s$ and $1, q, q^2, q^3, \ldots, q^t$, as well as all of the possible product pairs, giving us a list of all possible divisors, added to each other.

Therefore, to find the sum of all the divisors of a number, we prime factorize that number, say, to $p^s q^t r^u$, and then the sum of all the divisors is given by

$$\left(1+p+p^2+p^3+\ldots+p^s\right)\left(1+q+q^2+q^3+\ldots+q^t\right)\left(1+r+r^2+r^3+\ldots+r^u\right).$$

Let's try it for 60; its divisors are 1, 2, 3, 4, 5, 6, 10, 12, 15, 20, 30, 60. This sums to 168 (remember, we're now looking at all the divisors, including 60 itself). The prime factorization of 60 is $5 \cdot 2^2 \cdot 3$. Therefore, we expect that the sum of the divisors will be given by the product $(1 + 5)(1 + 2 + 2^2)(1 + 3) = (6)(7)(4) = 168$. This checks out. If 60 were a perfect number, the above product would have equaled $2N$, or 120. So we know 60 is not a perfect number.

If we check for 28, we get that $28 = 7 \cdot 2^2$. The sum of its divisors is thus $(1 + 7)(1 + 2 + 2^2) = (8)(7) = 56$, which is $2 \cdot 28$, so we confirm that 28 is a perfect number.

This approach gives us a shortcut, but it is still quite lengthy because we have to go through and prime factorize each of the numbers, form the products as above, and so on. Let's find another approach, perhaps more streamlined.

Insight 2

The above approach gave us two important ideas. First, when we look at divisors, we should look at them in terms of prime factorization and a product of the form

$$\left(1+p+p^2+p^3+\ldots+p^s\right)\left(1+q+q^2+q^3+\ldots+q^t\right).$$

Second, to check whether a number is a perfect number, it is more natural to think about summing *all* of the divisors of a number N to see whether the sum comes out to $2N$.

Let's look at the two perfect numbers we have so far: $6 = 2 \times 3$ and $28 = 2^2 \times 7$. With a bit of ingenuity, we can rewrite the factors as

$$6 = 2 \times 3 = 2^1\left(2^2 - 1\right),$$

$$28 = 2^2 \times 7 = 2^2\left(2^3 - 1\right).$$

When we write the numbers this way, all of a sudden we feel that we are on to something. We see a pattern. Each number is written as $N = 2^{p-1}(2^p - 1)$, where $2^p - 1$ is prime (such as 3 when $p = 2$, or 7 when $p = 3$). This is the sort of insight that we've called type 3 insight, where from a few cases, we see a pattern that we can try to extend—we guess that this pattern is significant, and try to prove it.

So we hypothesize that if a number takes the form $N = 2^{p-1}(2^p - 1)$, where $2^p - 1$ is prime, then N is a perfect number.

From our work above, we know that the sum of all of the divisors can be written out as a sum of the form

$$\left(1+2+2^2+\ldots+2^{p-1}\right)\left(1+\left(2^p-1\right)\right), \qquad (3.17)$$

precisely analogous to $(1 + 2 + 2^2)(1 + 7)$ for the case of 28. We see that the right-hand factor of Equation (3.17), $(1 + (2^p - 1))$, is just 2^p. Let's look at the left-hand factor, $(1 + 2 + 2^2 + \ldots + 2^{p-1})$. This is just the sum of a geometric series, with common ratio 2 and last term 2^{p-1}. From Chapter 1, we know that the sum of such a series is

$$\frac{1-2\cdot 2^{p-1}}{1-2} = 2^p - 1.$$

Therefore, the sum of all the divisors of N is$2^p(2^p - 1)$, which can be rewritten as $2[2^{p-1}(2^p - 1)]$. Now we recall that, by our assumption, we have $N = 2^{p-1}(2^p - 1)$, so the sum of the divisors is $2N$.

Thus, we have proven that if N is of the form $N = 2^{p-1}(2^p - 1)$, where $2^p - 1$ is prime, then N is a perfect number. We also see why $2^p - 1$ needs to be prime—because we need its factors to be just 1 and $2^p - 1$, so that the right-hand factor of Equation (3.17) is just $(1 + (2^p - 1))$. Therefore, we have transformed our problem of finding perfect numbers into the problem of finding numbers for which $(2^p - 1)$ is prime. If we find such a prime, then we have found ourselves a perfect number $N = 2^{p-1}(2^p - 1)$.

Looking at our list so far,

$$6 = 2\cdot 3 = 2^1\left(2^2-1\right),$$

$$28 = 2^2\cdot 7 = 2^2\left(2^3-1\right),$$

we guess that the next perfect number would be $N = 2^3(2^4 - 1) = 8(15) = 120$. Its divisors are 1, 2, 3, 4, 5, 6, 10, 12, 15, 20, 30, 40, 60, 120. This sums to 328, but that is not equal to 2(120). What happened? We forgot that $2^p - 1$

has to be prime, and $2^4 - 1 = 15$, which is definitely not prime. That is our problem. Therefore, we check the next possibility, $2^5 - 1 = 31$. This is prime. Therefore, we know that $2^4(2^5 - 1) = 16(31) = 496$ is a perfect number, and we have solved our problem.

A few interesting points should be made here, just to broaden our horizons. First, we can see how much time this approach saves. Imagine having to check all of the numbers from 28 to 496, writing them in a big table, finding all their divisors, and then finding the sum of the divisors. Even using our shortcut of prime factorization from Insight 1, this would be an unbelievable amount of work.

Second—and this is a subtle point—we proved that if $N = 2^{p-1}(2^p - 1)$, where $2^p - 1$ is prime, then N is a perfect number. However, this is not the same as proving that all perfect numbers are of this form. Therefore, although 496 is the next perfect number of this form after 28, perhaps there is a smaller perfect number of a different form. That is why the problem above was phrased, "Can we find another perfect number?" and not "Can we find the next perfect number?"

Our result was known to Euclid and the ancient Greeks, but it had to wait for Euler, nearly 2,000 years later, to prove that all even perfect numbers have this form. That is, an even number is perfect *if and only if* $N = 2^{p-1}(2^p - 1)$, where $2^p - 1$ is prime. Therefore, we can feel confident that 496 is indeed the next even perfect number. This raises the question of whether there are any odd perfect numbers. So far, none have been discovered, and if one exists, mathematicians have proven that it would be extremely large (greater than 10^{300}).

Third, we see that the search for perfect numbers is then really the search for primes of the form $2^p - 1$. These are called Mersenne primes, after the 17th century monk, Marin Mersenne, who studied number theory. Whether there are an infinite or finite number of Mersenne primes (and hence perfect numbers) is unknown. As of June 2010, only 47 Mersenne primes, and therefore 47 even perfect numbers, are known. The largest of these is $2^{43,112,608} \times (2^{43,112,609} - 1)$ with 25,956,377 digits.

Last, we used conjecture to get an insight. This is useful but sometimes dangerous. The ancient Greeks knew of only four perfect numbers:

$$P_1 = 6 = 2^1\,(2^2 - 1),$$

$$P_2 = 28 = 2^2\,(2^3 - 1),$$

$$P_3 = 496 = 2^4\,(2^5 - 1),$$

$$P_4 = 8{,}128 = 2^6\,(2^7 - 1).$$

Looking at patterns, as we did, mathematicians of the Middle Ages conjectured that there was probably one perfect number between each power of 10 and therefore that the *n*th perfect number would be *n* digits long, and that perfect numbers would always end alternately in 6 or 8. Therefore, the fifth perfect number should have five digits and end in a 6. In this case, both conjectures were incorrect. In fact, $P_5 = 2^{12}(2^{13} - 1) = 33{,}550{,}336$, so there is no perfect number of five digits (although it did end in 6). Also, $P_6 = 2^{16}(2^{17} - 1) = 8{,}589{,}869{,}056$, also ending in 6. Therefore, insights based on conjecture are good, but we have to be careful that they are not construed as proofs.

Tools Used and Developed

- Prime factorization.
- An even number N is perfect if and only if $N = 2^{p-1}(2^p - 1)$, where $2^p - 1$ is prime.

Problem 3.7

Samir's mother asked him how many students were in his seventh-grade class. He couldn't remember, but said, "Throughout the year we were divided into groups for projects. I know that when we were divided into groups of three, there were two people left over. When were divided into groups of five, three people were left over, and when we were divided into groups of seven, four people were left over." What is the minimum number of students in Samir's class?

This type of problem is easy. We just need to check though the integers to find the first integer that leaves a remainder of 2 when divided by 3, a remainder of 3 when divided by 5, and a remainder of 4 when divided by 7.

We could make a table such as the following one, and check systematically.

Number	*Remainder for 3*	*Remainder for 5*	*Remainder for 7*	*Number*	*Remainder for 3*	*Remainder for 5*	*Remainder for 7*
1	1	1	1	28	1	3	0
2	2	2	2	29	2	4	1
3	0	3	3	30	0	0	2
4	1	4	4	31	1	1	3
5	2	0	5	32	2	2	4
6	0	1	6	33	0	3	5
7	1	2	0	34	1	4	6
8	2	3	1	35	2	0	0
9	0	4	2	36	0	1	1
10	1	0	3	37	1	2	2
11	2	1	4	38	2	3	3
12	0	2	5	39	0	4	4
13	1	3	6	40	1	0	5
14	2	4	0	41	2	1	6
15	0	0	1	42	0	2	0
16	1	1	2	43	1	3	1
17	2	2	3	44	2	4	2
18	0	3	4	45	0	0	3
19	1	4	5	46	1	1	4
20	2	0	6	47	2	2	5
21	0	1	0	48	0	3	6
22	1	2	1	49	1	4	0
23	2	3	2	50	2	0	1
24	0	4	3	51	0	1	2
25	1	0	4	52	1	2	3
26	2	1	5	**53**	**2**	**3**	**4**
27	0	2	6	54	0	4	5

By scanning down each column of this table, we can see that the remainders cycle in an orderly fashion, but it takes a long time to find just the right combination. Finally, 53 gives us the required results: a remainder of 2 when divided by 3, a remainder of 3 when divided by 5, and a remainder of 4 when divided by 7. Therefore, the smallest number of students in Samir's class is 53. (Samir is clearly a victim of educational budget cutbacks and growing classroom sizes!)

We could, of course, have made the problem easier by just checking all the numbers that leave a remainder of 3 when divided by 5 (numbers of the form $n = 5k + 3$ for all integers k), such as 8, 13, 18, ..., because these are easy to generate mentally, and then we could check their remainders when divided by 2 and 7. This is a nice shortcut, but the process is still rather tedious.

Suppose we now want to find the least number of students in the entire seventh grade, for which we know the following. Like Samir's class, the remainder for division by 3 is 2, the remainder for division by 5 is 3, the remainder for division by 7 is 4, but also the remainder for division by 11 is 5.

For this, we need an insight. Otherwise, it would be way too tedious. This sort of insight is much easier to come by if we have already studied some math, particularly a branch called number theory, but it's okay if we haven't. It is still worth thinking about the problem to find a smart solution.

Insight

We know that the problem focuses on remainders. We see from the table above that when we look at the remainders for division by 3, the numbers cycle, leaving remainders of 1, 2, 0, and then 1, 2, 0, again, and so on. The same sort of thing happens when we look at division by 5 or 7. We see, in fact, that for any integer n, the natural numbers can be divided into classes by what remainder they leave when divided by n. The possible remainders are $0, 1, 2, \ldots, n - 1$.

The mathematical system developed to think about numbers in this way is known as *modular* arithmetic. It is actually not too difficult. The notation to say that 7 leaves a remainder of 2 when divided by 5 is $7 \equiv 2 \pmod 5$, which would be read as, "seven is congruent to two, modulo five." The strange-looking "equals" sign with three bars is called a congruence, and in this context, it means that 7 is one of the numbers in the class that leaves a remainder of 2 when divided by 5 (i.e., in terms of remainders, it is equivalent to 2). Other true statements would be $12 \equiv 2 \pmod 5$, $23 \equiv 3 \pmod 5$, and $40 \equiv 0 \pmod 5$.

We see, in terms of remainders with respect to 5, that 7 and 12 are equivalent because they both leave the same remainder. Thus, we could also say $12 \equiv 7 \pmod 5$

We can quickly deduce several interesting properties of modular arithmetic. If we say $a \equiv b \pmod m$, this implies that $a - b$ is divisible by m, or that b differs from a by a multiple of m, or $b = a + km$. This is because if $a \equiv b \pmod m$, then both a and b leave the same remainder r when divided by m. Thus, we can write $a = sm + r$, and $b = tm + r$ for some integers s and t. Thus, a and b both have remainder r when divided by m because they are r more than some multiple of m. Thus, $a - b = (sm + r) - (tm + r) = (s - t)m$, so $a - b$ is a multiple of m; that is, $a - b$ is divisible by m.

This leads to some interesting ideas, such as introducing negative numbers into congruences. We can thus write $-1 \equiv 4 \pmod{5}$ because -1 and 4 differ by a multiple of 5, so their difference $(4 - (-1))$ is divisible by 5.

Another interesting application is to explore operations with congruences, such as addition and multiplication. For example, we can deduce that if $a \equiv b \pmod{m}$ and $c \equiv d \pmod{m}$, then

$$(a + c) \equiv (b + d) \pmod{m}.$$

We can prove this easily by remembering that a definition of congruence is that the difference between the two integers is divisible by m. Thus, we note that the difference

$$(a + c) - (b + d) = (a - b) + (c - d).$$

Because $a \equiv b \pmod{m}$ and $c \equiv d \pmod{m}$, then both $(a - b)$ and $(c - d)$ are divisible by m, and so is their sum. Also, if $a \equiv b \pmod{m}$ and $c \equiv d \pmod{m}$, then

$$ac \equiv bd \pmod{m}.$$

This is a touch harder to prove, but not too difficult. By using our notation from before, we can write $a = sm + r_1$ and $b = tm + r_1$; that is, both have the same remainder r_1 when divided by m. Similarly, we can write $c = jm + r_2$ and $d = km + r_2$, meaning that they both have a different remainder r_2 when divided by m.

If we multiply, we get

$$ac = sjm^2 + smr_2 + jmr_1 + r_1r_2,$$

$$bd = tkm^2 + tmr_2 + kmr_1 + r_1r_2.$$

We see that in each product, the first three terms are divisible by m, and that for both products, the remainder is the same, r_1r_2. We can check this with an example, such as $7 \equiv 2 \pmod{5}$ and $8 \equiv 3 \pmod{5}$. Thus, we expect that $7 \times 8 \equiv 2 \times 3 \pmod{5}$, or $56 \equiv 6 \pmod{5}$. This is true, since both leave a remainder of 1 when divided by 5.

This sort of insight becomes very useful when we note that if we multiply a by something congruent to 1 (mod m), we do not change the congruence of a with respect to m. Thus, if we know that $7 \equiv 2 \pmod 5$, we also immediately know that $7 \times 2{,}011 \equiv 2 \pmod 5$. That is because $2{,}011 \equiv 1 \pmod 5$, so $7 \times 2{,}011 \equiv 2 \times 1 \pmod 5$. We can check directly: $7 \times 2{,}011 = 14{,}077 = (2{,}815 \times 5) + 2$.

Armed with these facts, we can now tackle our seventh-grade class size problem in an insightful way. We have a problem of the following type: Find a number x that satisfies multiple congruences simultaneously. In our case,

$$x \equiv 2 \pmod 3, \tag{3.18a}$$

$$x \equiv 3 \pmod 5, \tag{3.18b}$$

$$x \equiv 4 \pmod 7, \tag{3.18c}$$

$$x \equiv 5 \pmod{11}. \tag{3.18d}$$

We notice that all the moduli, 3, 5, 7, and 11, are prime numbers, which makes it easier. The following analysis just depends on the modulo being relatively prime to each other (i.e., having no common factors), a condition certainly satisfied here. Also, we notice that for each equation of the form, $x \equiv a_i \pmod{m_i}$, each a_i is relatively prime to its corresponding m_i. In other words, in Equation (3.18a), 2 has no common factors with 3, in Equation (3.18b), 3 has no common factors with 5, and so on. Momentarily, we will see why these circumstances are so special.

Let us solve our problem generally,

$$\begin{aligned} x &\equiv a_1 \pmod{m_1}, \\ x &\equiv a_2 \pmod{m_2}, \\ x &\equiv a_3 \pmod{m_3}, \\ &\vdots \\ x &\equiv a_i \pmod{m_i}, \\ &\vdots \\ x &\equiv a_n \pmod{m_n}. \end{aligned} \tag{3.19}$$

Here, we know by assumption that $m_1, m_2, m_3, \ldots, m_n$ are relatively prime. Thus, we decide that we will make one large modulo, being the product of all the smaller moduli, or

$$M = m_1 m_2 m_3 \ldots m_n.$$

We know that when M is divided by m_1,

$$\frac{M}{m_1} = m_2 m_3 \ldots m_n,$$

the quotient is divisible by all the other modulo, $m_2, m_3, \ldots, m_n$, and is relatively prime to m_1. This holds true across the board. The quotient

$$\frac{M}{m_i} = m_1 m_2 \ldots m_{i-1} m_{i+1} \ldots m_n$$

is divisible by all modulo except m_i, to which it is relatively prime.

Here is the key insight. If we can find numbers b_i that solve the very simple congruences

$$b_i \frac{M}{m_i} \equiv 1 \pmod{m_i}, \tag{3.20}$$

which should be fairly easy, since we are seeking a simple congruence to 1(mod m_i), then our desired number x is generated as follows. We want a number x that will satisfy the following congruence:

$$x \equiv \left(a_1 b_1 \frac{M}{m_1} + a_2 b_2 \frac{M}{m_2} + \ldots + a_n b_n \frac{M}{m_n} \right) (\text{mod } M). \tag{3.21}$$

If we can generate this x, we see, for example, that x also satisfies the first congruence relation, $x \equiv a_1 \pmod{m_1}$ because, after the first term, all terms of the form $a_i b_i (M/m_i)$ are divisible by m_1. This is because, for $i \neq 1$, $M/m_i = m_1 m_2 \ldots m_{i-1} m_{i+1} \ldots m_n$, and this contains an m_1. Therefore, $x \equiv [a_i b_i (M/m_i)] \pmod{m_1}$. However, we have already (by assumption) found a b_1 such that $b_1(M/m_i) \equiv 1 \pmod{m_1}$. Therefore,

$$\left(a_1 b_1 \frac{M}{m_1} \right) \equiv a_1 \times 1 \pmod{m_1} \equiv a_1 \pmod{m_1}.$$

The same argument holds for each m_i, so we see that our number x generated by Equation (3.21) will automatically satisfy the system of congruences specified in Equation (3.19). To show how this works, let us now go back to our system specified in Equations (3.18a–3.18d):

$$x \equiv 2 \pmod{3},$$

$$x \equiv 3 \pmod{5},$$

$$x \equiv 4 \pmod{7},$$

$$x \equiv 5 \pmod{11}.$$

Here, $a_1 = 2$, $m_1 = 3$, $a_2 = 3$, $m_2 = 5$, $a_3 = 4$, $m_3 = 7$, $a_4 = 5$, $m_4 = 11$. Therefore,

$$M = m_1 m_2 m_3 m_4 = 3 \times 5 \times 7 \times 11 = 1{,}155,$$

and

$$\frac{M}{m_1} = 385, \frac{M}{m_2} = 231, \frac{M}{m_3} = 165, \frac{M}{m_4} = 105.$$

Therefore, we are seeking numbers b_i that solve the very simple congruences $b_i(M/m_i) \equiv 1 \pmod{m_i}$, so we want to solve the following system:

$$385b_1 \equiv 1 \pmod{3}, 231b_2 \equiv 1 \pmod{5}, 165b_3 \equiv 1 \pmod{7}, 105b_4 \equiv 1 \pmod{11}.$$

Here it is easy to see that $b_1 = 1$ because $(3 + 8 + 5) = 16$, which is 1 more than 15, the digit sum of 384, so 384 is divisible by 3). Also, we see that $b_2 = 1$ directly, since 231 is 1 more than 230, which is obviously divisible by 5. By quick trial and error, we see that $b_3 = 2$, since 330 has a remainder of 1 when divided by 7. Also, we see quickly that $b_4 = 2$, since $210 = (19 \times 11) + 1$. So, plugging these values into Equation (3.21), we see that

$$x \equiv (2 \times 1 \times 385 + 3 \times 1 \times 231 + 4 \times 2 \times 165 + 5 \times 2 \times 105) \pmod{1{,}155},$$

or

$$x \equiv 3{,}833 \pmod{1{,}155}.$$

We see that 1,155 can go into 3,833 three times, leaving a remainder of 368. Therefore, $x \equiv 368 \pmod{1{,}155}$, and that is the smallest value we can get. So, there are 368 students in the entire seventh grade. We can check quickly to see that this satisfies the congruences:

$368 = (122 \times 3) + 2$, $368 = (73 \times 5) + 3$, $368 = (52 \times 7) + 4$, $368 = (33 \times 11) + 5$.

Now that we have gained this insight, let's see how quickly we could have solved our original problem:

$$x \equiv 2 \pmod{3},$$

$$x \equiv 3 \pmod{5},$$

$$x \equiv 4 \pmod{7}.$$

Here, $a_1 = 2$, $m_1 = 3$, $a_2 = 3$, $m_2 = 5$, $a_3 = 4$, $m_3 = 7$. Therefore, $M = m_1m_2m_3 = 3 \times 5 \times 7 = 105$, and $M/m_1 = 35$, $M/m_2 = 21$, $M/m_3 = 15$.

So we are seeking numbers b_i that solve the very simple congruences $b_i(M/m_i) \equiv 1 \pmod{m_i}$, and we want to solve the following system:

$$35b_1 \equiv 1 \pmod{3}, \quad 21b_2 \equiv 1 \pmod{5}, \quad 15b_3 \equiv 1 \pmod{7}.$$

This system, by very quick trial and error, has the solutions $b_1 = 2$, $b_2 = 1$, $b_3 = 1$, so then Equation (3.21) becomes: $x \equiv (2 \times 2 \times 35 + 3 \times 1 \times 21 + 4 \times 1 \times 15) \pmod{105}$, or $x \equiv 263 \pmod{105}$. We can see that 105 goes into 263 twice (giving 210) with a remainder of 53. Therefore, $x \equiv 53 \pmod{105}$, giving us our first answer much faster than going through and checking numbers.

The procedure for solving multiple congruences that we used above is often called the Chinese remainder theorem, as the procedure was apparently known to the ancient Chinese. However, the invention and development of modular arithmetic as a full mathematical system (a theory of congruences) is due to—guess who? None other than Gauss himself, in a book often considered his masterpiece, *Disquisitiones arithemeticae*. So, even though this problem was a bit involved, it couldn't be left out of a book dedicated to Gaussian-type insights.

Tools Used and Developed

- Modular arithmetic.
- Chinese remainder theorem.

4

Algebra

Problem 4.1

There are two numbers whose sum is 7 and whose product is 3. Find the sum of their reciprocals.

This problem sounds straightforward. We set up the equations and solve. Naturally, we'll call the numbers x and y, and we have our two equations:

$$x + y = 7,$$

$$xy = 3.$$

We can substitute for y in the second equation, noting that the first equation gives us $y = 7 - x$. So, we have $x(7 - x) = 3$, or $7x - x^2 = 3$. This gives us the quadratic equation

$$x^2 - 7x + 3 = 0.$$

Now it's smooth sailing. We just bring in the quadratic formula,

$$x = \frac{7 \pm \sqrt{7^2 - 4(3)(1)}}{2} = \frac{7 \pm \sqrt{37}}{2}.$$

So we get two values for x, but actually, they are values for x and y because $y = 7 - x$, so if we take $x = (7+\sqrt{37})/2$, then $y = 7-(7+\sqrt{37})/2 = (7-\sqrt{37})/2$, and if we take $x = (7-\sqrt{37})/2$, then $y = (7+\sqrt{37})/2$.

We just need to take the reciprocals of these two numbers now, and sum them:

$$\frac{2}{7+\sqrt{37}} + \frac{2}{7-\sqrt{37}}.$$

Rationalizing the denominators and simplifying, we get

$$\frac{2}{7+\sqrt{37}}\frac{(7-\sqrt{37})}{(7-\sqrt{37})} + \frac{2}{7-\sqrt{37}}\frac{(7+\sqrt{37})}{(7+\sqrt{37})}$$

$$= \frac{14-2\sqrt{37}+14+2\sqrt{37}}{7^2-37} = \frac{28}{12} = \frac{7}{3}.$$

That took a bit of tedious work. Is there a faster way to do this problem? Yes, there is, and this calls for a subtle insight. We have to keep our eye on the ball.

Insight

We solved this problem by finding the two numbers, taking their reciprocals, and then adding. However, we were not asked to find the two numbers themselves, only the sum of their reciprocals. Is there a way to jump to that more directly?

Let's look at a trick that, once we've seen it, we'll always remember it and be able to pull it out. If we have two numbers, x and y, along with their sum and their product, we can set up the simple ratio $(x + y)/xy$ and then break up the fraction to get

$$\frac{x+y}{xy} = \frac{x}{xy} + \frac{y}{xy} = \frac{1}{y} + \frac{1}{x}.$$

We can do this last cancellation because we know that neither x nor y is 0, as their product is nonzero. Therefore, the sum of the reciprocals of x and y is just their sum divided by their product.

Going back to the information we had,

$$x + y = 7,$$

$$xy = 3,$$

we see that we can get $1/x + 1/y$ in a single step, as $(x + y)/xy = 7/3$. Pretty slick, and it saves a lot of work.

Tools Used and Developed

- Quadratic formula.
- $\dfrac{x+y}{xy} = \dfrac{1}{x} + \dfrac{1}{y}$.
- Keeping our eye on the ball—for example, if we're asked to calculate the sum of the reciprocals of two numbers, we may not have to find the two numbers first.

Problem 4.2

Let p, q, and r, be the solutions to the equation $2x^3 + 11x^2 + 22x + 15 = 0$.

a. Find $p + q + r$.
b. Find $p^2 + q^2 + r^2$.
c. Find $\dfrac{1}{p} + \dfrac{1}{q} + \dfrac{1}{r}$.

Many algebraic problems require either finding the roots of polynomials, or even performing operations with these roots. Algebra I courses teach that the roots of $ax^2 + bx + c = 0$ are

$$x = \frac{-b \pm \sqrt{b^2 - 4ac}}{2a}.$$

Thus, to find the sum of the roots, let's call them r and s, we could use the quadratic formula to find both r and s, and then simply add $r + s$. Likewise, we could get the product of the roots, rs, or even $r^2 + s^2$, or anything of the sort.

But how should we approach this cubic polynomial problem? One option is to use the formula for finding the roots of cubic polynomials, which states that for equations of the form $ax^3 + bx^2 + cx + d = 0$, the roots are

$$x = \sqrt[3]{\left(\frac{-b^3}{27a^3} + \frac{bc}{6a^2} - \frac{d}{2a}\right) + \sqrt{\left(\frac{-b^3}{27a^3} + \frac{bc}{6a^2} - \frac{d}{2a}\right)^2 + \left(\frac{c}{3a} - \frac{b^2}{9a^2}\right)^3}}$$

$$+ \sqrt[3]{\left(\frac{-b^3}{27a^3} + \frac{bc}{6a^2} - \frac{d}{2a}\right) - \sqrt{\left(\frac{-b^3}{27a^3} + \frac{bc}{6a^2} - \frac{d}{2a}\right)^2 + \left(\frac{c}{3a} - \frac{b^2}{9a^2}\right)^3}} - \frac{b}{3a}.$$

So, we could try to use $a = 2$, $b = 11$, $c = 22$, $d = 15$ to solve for p, q, and r and answer the three parts of our problem, but this looks incredibly painful, tedious, and—let's face it—boring.

Our other option is to use the rational root theorem, which states that for equations of the form $a_n x^n + a_{n-1}x^{n-1} + \ldots + a_0 = 0$, each rational solution x will be equal to $x = m/n$, where m is an integer factor of a_0 and n is an integer factor of a_n. So, for our equation, $2x^3 + 11x^2 + 22x + 15 = 0$, we know that our rational solution x will fall in the form of $x = m/n$ where m can equal ± 1, ± 3, ± 5, ± 15, and n can equal ± 1, ± 2. We can garner some context clues; for example, since all of the terms of $2x^3 + 11x^2 + 22x + 15$ are positive, any rational root must be negative, thereby eliminating some possibilities. If we were to try out our various possible fractions to solve for x (a very time consuming process), we would find that $-3/2$ is, in fact, a root of our polynomial. This means that $(2x + 3)$ is a factor of our polynomial. We could then do some long division or synthetic division to find that our polynomial $2x^3 + 11x^2 + 22x + 15$ can be factored into $(2x + 3)(x^2 + 4x + 5)$, and we could then solve for the remaining roots by using the quadratic formula. The roots are $p = -3/2$, $q = -2 + 2i$, and $r = -2 - 2i$. We can then use these roots to find $p + q + r$, $p^2 + q^2 + r^2$, and $1/p + 1/q + 1/r$.

Phew—that was hard! Clearly, we need an insight into a more efficient way to solve such problems.

Insight

Luckily, our insight comes in the form of Vieta's formula. Like the quadratic equation, Vieta's formula proposes a relationship between the coefficients of a polynomial and its roots. Let's first try to understand Vieta's formula in relation to quadratic equations, and then we will apply it to higher-order polynomials.

We know that a quadratic equation of the form $ax^2 + bx + c$, can always be factored into $a(x - r)(x - s)$, where r and s are the roots of the equation. Hence, we can rewrite our quadratic equation as

$$ax^2 + bx + c = a(x - r)(x - s). \tag{4.1}$$

If we expand the right-hand side of Equation (4.1), we get

$$ax^2 + bx + c = ax^2 - a(r + s)x + ars.$$

If we now equate the coefficients of each power of x, we see that

$$a = a,$$

$$b = -a(r + s),$$

$$c = ars.$$

Thus, we have related the a, b, and c coefficients to the roots r and s.

So to solve for $r + s$, we could easily use the formula $b = -a(r + s)$, to see that $r + s = -b/a$. Likewise, to solve for rs, we would use the formula $c = ars$ to find that $rs = c/a$. By using these manipulations, we can now solve many similar problems. Let's now try to apply this logic to our cubic polynomial.

We know that a cubic polynomial of the form $ax^3 + bx^2 + cx + d$ can be factored into $a(x - p)(x - q)(x - r)$, where p, q, and r are the roots. Hence, we have

$$ax^3 + bx^2 + cx + d = a(x - p)(x - q)(x - r). \tag{4.2}$$

Expanding the right-hand side of Equation (4.2) yields

$$ax^3 + bx^2 + cx + d = ax^3 - a(p + q + r)x^2 + a(pq + qr + rp)x - apqr.$$

This means that the coefficients are now related to the roots through the following equations:

$$a = a, \tag{4.3}$$

$$b = -a(p + q + r), \tag{4.4}$$

$$c = a(pq + qr + rp), \tag{4.5}$$

$$d = -apqr. \tag{4.6}$$

We can now use this information to answer the original three-part problem.

For part (a), we see from Equation (4.4) that $b = -a(p + q + r)$; hence, $(p + q + r) = -b/a = -11/2$.

For part (b), we already know $(p + q + r)$ from part (a), so our first step should be to square this in order to get the terms, p^2, q^2, and r^2. Thus,

$$(p+q+r)^2 = \left(-\frac{11}{2}\right)^2,$$

$$p^2 + q^2 + r^2 + 2pq + 2qr + 2pr = \frac{121}{4},$$

$$p^2 + q^2 + r^2 = \frac{121}{4} - 2(pq + qr + pr).$$

We know from Equation (4.5) that $c = a(pq + qr + rp)$; hence, $(pq + qr + rp) = c/a$. So,

$$p^2 + q^2 + r^2 = \frac{121}{4} - 2\left(\frac{c}{a}\right) = \frac{121}{4} - 2\left(\frac{22}{2}\right) = \frac{121}{4} - 22 = \frac{33}{4}.$$

Finally, for part (c), to find $1/p + 1/q + 1/r$, we should write these fractions with a common denominator:

$$\frac{1}{p} + \frac{1}{q} + \frac{1}{r} = \frac{pq + qr + pr}{pqr}.$$

Once again, we know from Equation (4.5) that $c = a(pq + qr + rp)$, so $(pq + qr + rp) = c/a$. Likewise, we know from Equation (4.6) that $d = -apqr$, so $pqr = -d/a$. Thus, we have

$$\frac{1}{p} + \frac{1}{q} + \frac{1}{r} = \frac{pq + qr + pr}{pqr} = \frac{\frac{c}{a}}{\left(\frac{-d}{a}\right)} = \frac{c}{-d} = -\frac{22}{15}.$$

With Vieta's formula, we can relate the coefficients of any polynomial to its roots and perform all sorts of operations.

Tools Used and Developed

- How to relate the coefficients of polynomials to their roots through various formulas, and how to perform operations with these roots.
- Vieta's formula, used to manipulate the roots of polynomials knowing only the coefficients of the variables.

Problem 4.3

Let r be a root of $x^2 - 2x + 3$. Find the value of $r^4 - 4r^3 + 2r^2 + 4r + 3$.

When working with quadratic equations, it is often not enough to just be able to figure out the roots. Advanced mathematicians are not only able to manipulate the roots, they are able to perform operations without even figuring out what the roots are. This is one such problem.

Of course, we can use the quadratic formula to figure out the roots of $x^2 - 2x + 3$, and then we can perform the indicated operations:

$$x = \frac{-b \pm \sqrt{b^2 - 4ac}}{2a} = \frac{2 \pm \sqrt{(-2)^2 - (4)(1)(3)}}{2(1)}$$

$$= \frac{2 \pm \sqrt{4-12}}{2} = \frac{2 \pm \sqrt{-8}}{2} = \frac{2 \pm 2i\sqrt{2}}{2} = 1 \pm \sqrt{2}i.$$

So we know that the two roots of our quadratic equation are $x = 1 + \sqrt{2}i$ and $x = 1 - \sqrt{2}i$. We can now use these to figure out the value of $r^4 - 4r^3 + 2r^2 + 4r + 3$. For root $r = 1 + \sqrt{2}i$, we get

$$r^2 = (1+\sqrt{2}i)(1+\sqrt{2}i) = 2\sqrt{2}i - 1,$$

$$r^3 = (1+\sqrt{2}i)(2\sqrt{2}i - 1) = \sqrt{2}i - 5,$$

$$r^4 = (1+\sqrt{2}i)(\sqrt{2}i - 5) = -4\sqrt{2}i - 7.$$

We can now plug these values into our equation:

$$r^4 - 4r^3 + 2r^2 + 4r + 3$$
$$= (-4\sqrt{2}i - 7) - 4(\sqrt{2}i - 5) + 2(2\sqrt{2}i - 1) + 4(1 + \sqrt{2}i) + 3.$$

Clearly, this is going to get very messy, but let's just try to stay organized as we work and simplify one term at a time. Let's start by trying to consolidate the first two terms only:

$$(-4\sqrt{2}i - 7) - 4(\sqrt{2}i - 5) + [2(2\sqrt{2}i - 1) + 4(1 + \sqrt{2}i) + 3]$$
$$= -4\sqrt{2}i - 7 - 4\sqrt{2}i + 20 + [2(2\sqrt{2}i - 1) + 4(1 + \sqrt{2}i) + 3]$$
$$= -8\sqrt{2}i + 13 + [2(2\sqrt{2}i - 1) + 4(1 + \sqrt{2}i) + 3].$$

Now let's consolidate the next term with that result:

$$-8\sqrt{2}i + 13 + [2(2\sqrt{2}i - 1) + 4(1 + \sqrt{2}i) + 3]$$
$$= -8\sqrt{2}i + 13 + 4\sqrt{2}i - 2 + [4(1 + \sqrt{2}i) + 3]$$
$$= -4\sqrt{2}i + 11 + [4(1 + \sqrt{2}i) + 3].$$

And finally, let's take on the last terms:

$$-4\sqrt{2}i + 11 + [4(1 + \sqrt{2}i) + 3]$$
$$= 4\sqrt{2}i + 11 + 4 + 4\sqrt{2}i + 3$$
$$= 18.$$

So, we can now say that $r^4 - 4r^3 + 2r^2 + 4r + 3 = 18$ for $r = 1 + \sqrt{2}i$.

Now we must compute the same for $r = 1 - \sqrt{2}i$.

$$r^2 = (1 - \sqrt{2}i)(1 - \sqrt{2}i) = -2\sqrt{2}i - 1,$$
$$r^3 = (1 + \sqrt{2}i)(-2\sqrt{2}i - 1) = -\sqrt{2}i - 5,$$
$$r^4 = (1 + \sqrt{2}i)(\sqrt{2}i - 5) = 4\sqrt{2}i - 7.$$

We can now plug these values into our equation:

$$r^4 - 4r^3 + 2r^2 + 4r + 3$$
$$= (4\sqrt{2}i - 7) - 4(-\sqrt{2}i - 5) + 2(2\sqrt{2}i - 1) + 4(1 - \sqrt{2}i) + 3.$$

Let's simplify this just as we did before:

$$(4\sqrt{2}i - 7) - 4(-\sqrt{2}i - 5) + [2(-2\sqrt{2}i - 1) + 4(1 - \sqrt{2}i) + 3]$$
$$= 4\sqrt{2}i - 7 + 4\sqrt{2}i + 20 + [2(-2\sqrt{2}i - 1) + 4(1 - \sqrt{2}i) + 3]$$
$$= 8\sqrt{2}i + 13 + [2(-2\sqrt{2}i - 1) + 4(1 - \sqrt{2}i) + 3]$$
$$= 8\sqrt{2}i + 13 + 2(-2\sqrt{2}i - 1) + [4(1 - \sqrt{2}i) + 3]$$
$$= 8\sqrt{2}i + 13 - 4\sqrt{2}i - 2 + [4(1 - \sqrt{2}i) + 3]$$
$$= 4\sqrt{2}i + 11 + [4(1 - \sqrt{2}i) + 3]$$
$$= 4\sqrt{2}i + 11 + 4 - 4\sqrt{2}i + 3$$
$$= 18.$$

We see that $r^4 - 4r^3 + 2r^2 + 4r + 3 = 18$ also for the root $r = 1 - \sqrt{2}i$. Thus, we are able to say that for the root, r, of the quadratic equation $x^2 - 2x + 3 = 0$, the value of $r^4 - 4r^3 + 2r^2 + 4r + 3$ is 18.

Even though we were able to find a definitive answer for this problem, this was certainly tedious, and it is clear that were we asked to do another problem like this (or a polynomial of a higher order!), we would need an insight to make solving such problems more efficient.

Insight

The best insight comes in the realization that the easiest way to solve this problem is not to solve for the roots at all, but rather to continue to work with the root as the variable, r, and simplify our equation as such. Let's give this some more thought.

We know that if r is the root of $x^2 - 2x + 3$, then by definition this means that $r^2 - 2r + 3 = 0$. We can rewrite this as $r^2 = 2r - 3$, and already we have a term that we can substitute in for r^2. It is easy now to get terms for r^3 and r^4 simply by multiplying through by r. So if $r^2 = 2r - 3$, then $r^3 = 2r^2 - 3r$, and $r^4 = 2r^3 - 3r^2$. We now have a way to simplify nearly every term in our original equation of $r^4 - 4r^3 + 2r^2 + 4r + 3$. Let's start by substituting in for r^4, so that we can get rid of this term:

$$\begin{aligned} r^4 - 4r^3 + 2r^2 + 4r + 3 &= (2r^3 - 3r^2) - 4r^3 + 2r^2 + 4r + 3 \\ &= -2r^3 - r^2 + 4r + 3. \end{aligned}$$

We can substitute in for r^3 to get rid of that term:

$$\begin{aligned} -2r^3 - r^2 + 4r + 3 &= -2(2r^2 - 3r) - r^2 + 4r + 3 \\ &= -4r^2 + 6r - r^2 + 4r + 3 \\ &= -5r^2 + 10r + 3. \end{aligned}$$

Now let's substitute for r^2

$$\begin{aligned} -5r^2 + 10r + 3 &= -5(2r - 3) + 10r + 3 \\ &= -10r + 15 + 10r + 3 \\ &= 18. \end{aligned}$$

Again, we see that for the root r (without even having to solve for it!) the value of $r^4 - 4r^3 + 2r^2 + 4r + 3$ is 18.

Tools Used and Developed

- Manipulating the roots of quadratic equations without even having to solve for them.
- Performing operations with the complex roots of quadratic equations by using our knowledge of quadratics, exponents, and algebraic substitution.

Problem 4.4

Simplify $(i + 1)^{3200} - (i - 1)^{3200}$.

Many numbers have unique properties, but perhaps none so much as the imaginary number $i = \sqrt{-1}$. This problem clearly involves a lot of multiplication. Let's start off with a smaller problem just to get the hang of it: we'll simplify $(i + 1)^2 - (i - 1)^2$.

First, we calculate $(i + 1)^2 = (i + 1)(i + 1) = i^2 + 2i + 1 = 2i$. Likewise, $(i - 1)^2 = (i - 1)(i - 1) = i^2 - 2i + 1 = -2i$. Now we can solve $(i + 1)^2 - (i - 1)^2 = (2i) - (-2i) = 4i$.

Now let's solve $(i + 1)^3 - (i - 1)^3$. First, we calculate $(i + 1)^3 = (i + 1)^2 (i + 1) = (2i)(i + 1) = 2i^2 + 2i = 2i - 2$. Likewise, $(i - 1)^3 = (i - 1)^2 (i - 1) = (-2i)(i - 1) = -2i^2 + 2i = 2i + 2$. Now we can solve $(i + 1)^3 - (i - 1)^3 = (2i - 2) - (-2i + 2) = 4i - 4$.

As we can see, this is a very tedious process and we can't imagine performing these operations anymore, much less for $(i + 1)^{3200} - (i - 1)^{3200}$. Clearly, we need an insight.

Insight

Let's think back to what we said earlier about the unique properties of i. The imaginary number i is actually a great number to work with because the powers of i follow a fairly short and simple cycle. Let's take a look:

$$i^1 = i,$$

$$i^2 = (\sqrt{-1})(\sqrt{-1}) = -1,$$

$$i^3 = (-1)(\sqrt{-1}) = -i,$$

$$i^4 = (-i)(i) = 1.$$

As we can see, i^5 will be equal to i once again, and the powers of i will simply repeat through this cycle of four numbers (i.e., $i^5 = i$, $i^6 = -1$, $i^7 = -i$, etc.). This insight gives us a major breakthrough into our problem. If the powers of i cycle through these four products, perhaps the powers of $(i+1)$ follow a similar cycle. Let's find $(i+1)^4$ the long way to see whether our instincts are correct.

$$(i+1)^4 = (i+1)^3\,(i+1) = (2i-2)(i+1) = 2i^2 - 2 = -4.$$

This is great! Given that $(i+1)^4 = -4$, this gives us an easy number to work with when looking at the exponents of i. Likewise,

$$(i-1)^4 = (i-1)^3\,(i-1) = (2i+2)(i-1) = 2i^2 - 2 = -4.$$

Using what we know about exponents, we can rewrite our original problem, $(i+1)^{3200} - (i-1)^{3200}$, as

$$(i+1)^{3200} - (i-1)^{3200} = [(i+1)^4]^{800} - [(i-1)^4]^{800} = (-4)^{800} - (-4)^{800} = 0.$$

We have our solution! The most important thing to remember when working with imaginary numbers is the cycle of i.

Tools Used and Developed

- How the exponents of imaginary numbers function and the cycle that they follow.
- Using our knowledge of exponents to manipulate the exponents of imaginary numbers to solve problems.

Problem 4.5

Find the remainder when $2x^{10} - 3ix^8 + (1+i)x^2 - (3-2i)x + 1$ is divided by $x - i$.

Young math students learn how to do long division. To an elementary school student, long division seems like the hardest type of math there is. But for an algebra student, long division has become second nature and seems supremely easy, until polynomial division comes along.

We know that to solve this problem either some long division or synthetic division is necessary. Either way, we are working with complex coefficients, and the division is going to be rather lengthy. Personally, I prefer long division, as I believe directly seeing the variables and their order helps to ward off careless errors because I do not have to account for place values. Hence, we will set up this problem as

$$\begin{array}{l}
\qquad 2x^9 + 2ix^8 + (-3i-2)x^7 + (3-2i)x^6 + (3i+2)x^5 + (-3+2i)x^4 + (-3i-2)x^3 + (3-2i)x^2 + (3+4i)x + (-7+i) \\
x - i \overline{\big) \, 2x^{10} - 3ix^8 + (1+i)x^2 - (3+2i)x + 1.} \\
\underline{2x^{10} - 2ix^9} \\
2ix^9 - 3ix^8 \\
\underline{2ix^9 + 2x^8} \\
(-3i-2)x^8 \\
\underline{(-3i-2)x^8 - (3-2i)x^7} \\
(3-2i)x^7 \\
\underline{(3-2i)x^7 - (3i+2)x^6} \\
(3i+2)x^6 \\
\underline{(3i+2)x^6 - (-3+2i)x^5} \\
(-3+2i)x^5 \\
\underline{(-3+2i)x^5 - (-3i-2)x^4} \\
(-3i-2)x^4 \\
\underline{(-3i-2)x^4 - (3-2i)x^3} \\
(3-2i)x^3 + (1+i)x^2 \\
\underline{(3-2i)x^3 - (3i+2)x^2} \\
(3+4i)x^2 - (3+2i)x \\
\underline{(3+4i)x^2 - (3i-4)x} \\
(-7+i)x + 1 \\
\underline{(-7+i)x - (-7i-1)} \\
-7i
\end{array}$$

Wow—that was long! But we finally have our answer. When $2x^{10} - 3ix^8 + (1 + i)x^2 - (3 - 2i)x + 1$ is divided by $x - i$, the remainder is $-7i$. But clearly there must be a better system. We cannot possibly be expected to do this every time we are asked to divide polynomials, particularly if we are working with polynomials of higher order. Luckily, our insight comes in the form of the remainder theorem.

Insight

The remainder theorem states that when a polynomial $f(x)$ is divided by $x - a$, the remainder is $f(a)$. Let's try to understand why this is true.

When we divide a polynomial, $f(x)$, by a divisor, $x - a$, we expect to get a quotient, $q(x)$, and a remainder, r. We can express this as

$$f(x) = (x - a)q(x) + r, \tag{4.7}$$

which must hold true for all x. Since we are interested particularly in $f(a)$, we let $x = a$ in Equation (4.7). This gives

$$f(a) = (a - a)q(a) + r = 0 \times q(a) + r = r.$$

Thus, we see that when a polynomial $f(x)$ is divided by $(x - a)$, the remainder is $r = f(a)$.

Let's try our problem again with this information. We are looking to divide the polynomial $f(x) = 2x^{10} - 3ix^8 + (1 + i)x^2 - (3 - 2i)x + 1$ by $x - i$, so we would expect our remainder to be $f(i)$;

$$\begin{aligned} f(i) &= 2i^{10} - (3i)i^8 + (1+i)i^2 - (3-2i)i + 1 \\ &= 2(-1) - (3i)(1) + (1+i)(-1) - 3i - 2i^2 + 1 \\ &= -2 - 3i - 1 - i - 3i + 2 + 1 \\ &= -7i. \end{aligned}$$

We have our remainder! This was certainly much simpler than long division, thanks to the remainder theorem!

Tools Used and Developed

- Long division of polynomials.
- Using the remainder theorem to find the remainder of polynomial division problems.

Problem 4.6

Find x if $\sqrt[3]{6x+28} - \sqrt[3]{6x-28} = 2$.

This problem is a bit unusual. It is difficult and could literally take a few hours. We would guess that the best way to start is to isolate one of the radicals, $\sqrt[3]{6x+28} = 2+\sqrt[3]{6x-28}$, and cube both sides, which can be quite messy on the right-hand side of the equation. We use the formula that $(a+b)^3 = a^3+3a^2b+3ab^2+b^3$, where, in our case, $a = 2$ and $b = \sqrt[3]{6x-28}$. After some painful algebra, we get

$$6x+28 = 6x-20+12\sqrt[3]{6x-28}+6\sqrt[3]{(6x-28)^2}.$$

Simplifying this, we get

$$48 = 2\sqrt[3]{6x-28}+\sqrt[3]{(6x-28)^2}, \text{ or } 48-2\sqrt[3]{6x-28} = \sqrt[3]{(6x-28)^2}.$$

Once again, we proceed to cube both sides to get

$$-48x-13824\sqrt[3]{6x-28}+576\sqrt[3]{(6x-28)^2}+110816 = (6x-28)^2. \quad (4.8)$$

This looks like the solution would be very laborious and clearly is just making things worse. One thing we might try now is to substitute for $\sqrt[3]{(6x-28)^2}$, since we know that it equals $48-2\sqrt[3]{6x-28}$. If we do that on the left-hand side of Equation (4.8), we get

$$138464-48x-14976\sqrt[3]{6x-28} = (6x-28)^2.$$

That doesn't seem to have done too much, either. We've spent a lot of time expanding things, and our expression is still not simplified. Obviously, we need an insight.

Insight 1

We see that just cubing our expressions and trying to simplify doesn't work. So, let's start over with our expression, $\sqrt[3]{6x+28}-\sqrt[3]{6x-28} = 2$. Let's assign variables to stand for the complicated radicals, and see if we can find a way to get easier expressions. Let $a = \sqrt[3]{6x+28}$ and $b = \sqrt[3]{6x-28}$. Our expression can be written now as $a - b = 2$. That looks much simpler already. If we can find another way to relate a and b, maybe we can solve a system of equations, and then solve for x from there. We start thinking along these lines and look at getting rid of the radicals by cubing a and b individually:

$$a^3 = 6x + 28,$$

$$b^3 = 6x - 28.$$

We realize that we can subtract these two equations to get rid of x, and get a second equation in terms of a and b. So now we have the two equations

$$a - b = 2,$$

$$a^3 - b^3 = 56.$$

Things are looking much better. Now we have two obvious options to try to move further. We can factor $a^3 - b^3$, since we know that $a - b$ is one of its factors, or we can cube $a - b$ since we know that we'll get terms that include $a^3 - b^3$. We just need to explore the situation.

If we factor $a^3 - b^3 = 56$, we get $(a - b)(a^2 + ab + b^2) = 56$. However, we know that $a - b = 2$, so we have $a^2 + ab + b^2 = 28$. Looking at this, we see that if we substitute in $a = \sqrt[3]{6x+28}$ and $b = \sqrt[3]{6x-28}$, we'll still end up with rather messy terms. Rather than proceed down this road, we'll at least try our other strategy and see whether it looks easier. If not, we'll come back here and bash our heads against this expression.

Let's cube $a - b = 2$ to get $(a - b)^3 = a^3 - 3a^2b + 3ab^2 - b^3 = 8$. Since we know that $a^3 - b^3 = 56$, we can now say that $-3a^2b + 3ab^2 = -48$, or $a^2b - ab^2 = 16$. Since we know the value of $a - b$, we can factor this out and write $a^2b - ab^2 = ab(a - b) = 16$. Since $a - b = 2$, we now have $ab = 8$. This is looking great!

Now let's substitute our values $a = \sqrt[3]{6x+28}$ and $b = \sqrt[3]{6x-28}$ into this simple expression to get $\left(\sqrt[3]{6x+28}\right)\left(\sqrt[3]{6x-28}\right) = 8$. Combining the radicals, we get $\sqrt[3]{36x^2 - 784} = 8$. We finally see our way clear. We simply cube both sides to get $36x^2 - 784 = 512$. From here, we have $36x^2 = 1{,}296$, or $x^2 = 36$, giving us our final answer, $x = \pm 6$. We did it!

That was a great solution. We went from a seemingly endless cycle of cubing and expanding expressions to using variables to stand for our radicals, and manipulating them to get a solution. Nonetheless, we had to walk down a few different roads before we finally found our way. Let's look at another insight.

Insight 2

Our second insight is a type 2 insight that involves knowing a mathematical tool that we could bring in to help us directly crack this nut. The tool is the following theorem:

$$\text{If } a + b + c = 0, \text{ then } a^3 + b^3 + c^3 = 3abc. \tag{4.9}$$

This is a very general statement that can be brought to bear on a lot of different sorts of problems. We'll use it here shortly, but first, let's see where it comes from. A nice proof, which I reproduce here, is given in the excellent book *Mathematical Olympiad Treasures* by Titu Andreescu and Bogdan Enescu (Andreescu and Enescu, 2004).

Let us start by constructing a general cubic polynomial $P(x)$ whose roots are a, b, and c. We know that this polynomial can be written as $P(x) = (x - a)(x - b)(x - c)$. If we expand this out, we get

$$P(x) = x^3 - (a + b + c)x^2 + (ab + bc + ca)x - abc.$$

Since a, b, and c are roots of the polynomial, we know that $P(a) = P(b) = P(c) = 0$, and plugging in a, b, and c into $P(x)$ gives us the following equations:

$$a^3 - (a + b + c)a^2 + (ab + bc + ca)a - abc = 0,$$

$$b^3 - (a + b + c)b^2 + (ab + bc + ca)b - abc = 0,$$

$$c^3 - (a + b + c)c^2 + (ab + bc + ca)c - abc = 0.$$

If we add these three equations, we get

$$a^3 + b^3 + c^3 - (a + b + c)(a^2 + b^2 + c^2) + (ab + bc + ca)(a + b + c) - 3abc = 0.$$

From here, we see that if $a + b + c = 0$, the middle terms drop out, and we have $a^3 + b^3 + c^3 = 3abc$. Thus, we have proven the theorem given by Equation (4.9).

With this tool firmly in hand, let's return to finding x if $\sqrt[3]{6x+28}-\sqrt[3]{6x-28}=2$. We begin by rewriting this in a way that is amenable to our tool: $\sqrt[3]{6x+28}-\sqrt[3]{6x-28}-2=0$. Immediately, we see that we can assign our variables as $a=\sqrt[3]{6x+28}$, $b=-\sqrt[3]{6x-28}$, and $c = -2$, because now $a + b + c = 0$. Therefore, we know that $(\sqrt[3]{6x+28})^3+(-\sqrt[3]{6x-28})^3-8 = 3(\sqrt[3]{6x+28})(-\sqrt[3]{6x-28})(-2)$. Notice the negative sign in front of $\sqrt[3]{6x-28}$ as part of the definition of b. This is a bit different from our first solution, but it is what we need to make $a + b + c = 0$.

Simplifying this expression, we now have

$$(6x+28)-(6x-28)-8=3(\sqrt[3]{6x+28})(-\sqrt[3]{6x-28})(-2),$$

or $48=6(\sqrt[3]{36x^2-784})$, which leads to $8=\sqrt[3]{36x^2-784}$, taking us directly back to $36x^2 - 784 = 512$, or $x^2 = 36$, again giving us our answer, $x = \pm 6$.

This amazing tool made a very difficult problem trivial.

Tools Used and Developed

- Assigning variables to radicals to simplify complex equations. (In this case, we let $a=\sqrt[3]{6x+28}$ and $b=\sqrt[3]{6x-28}$, and then cubed $a - b$ and related that to $a^3 - b^3$.)
- If $a + b + c = 0$, then $a^3 + b^3 + c^3 = 3abc$.

Problem 4.7

Find x if

$$\frac{x^3-x^2+3x-4}{x^3-x^2-5x-5}=\frac{-x^3+x^2-3x+9}{-x^3+x^2+5x+10}.$$

We immediately start to cross multiply:

$$(x^3-x^2+3x-4)(-x^3+x^2+5x+10)$$
$$=(-x^3+x^2-3x+9)(x^3-x^2-5x-5).$$

We see that there seems to be some symmetry to the equation's terms and signs, but we don't quite know what to do with this, so we just start to cross multiply, knowing that we will get some cancellations.

The cross multiplications turn out to be extremely lengthy, and it takes great care to keep track of all the terms. At the end, we get

$$-x^6+2x^5+x^4+12x^3+x^2+10x-40$$

$$=-x^6+2x^5+x^4+12x^3+x^2-30x-45.$$

By crossing out terms and simplifying, we get $x = -1/8$.

This was truly time consuming. Is there an insight that could have helped us solve this problem more quickly and easily?

Insight

What we have here is a problem that involves the equality of two ratios. In the most general terms, we have something of the form $a/b = c/d$. When we have such an equality of ratios, there are some interesting results that we can derive and that can sometimes be very helpful. Let us start by calling the ratio r, such that $a/b = c/d = r$. Now we can say that $a = br$ and $c = dr$. Let's look at the ratio $(a + c)/(b + d)$, and substitute $a = br$ and $c = dr$ to get

$$\frac{br+dr}{b+d}=\frac{r(b+d)}{b+d}=r.$$

Therefore, we see that if $a/b = c/d$, then $a/b = (a + c)/(b + d)$.

Let's use this insight on our ratio of polynomials. Given that

$$\frac{x^3-x^2+3x-4}{x^3-x^2-5x-5}=\frac{-x^3+x^2-3x+9}{-x^3+x^2+5x+10},$$

we now also know that

$$\frac{x^3-x^2+3x-4}{x^3-x^2-5x-5}=\frac{(x^3-x^2+3x-4)+(-x^3+x^2-3x+9)}{(x^3-x^2-5x-5)+(-x^3+x^2+5x+10)}=\frac{5}{5}=1.$$

Therefore, we have $x^3 - x^2 + 3x - 4 = x^3 - x^2 - 5x - 5$, or $8x = -1$, and $x = -1/8$.

Thus, we can solve our problem very quickly and easily. This insight, that if $a/b = c/d$, then $a/b = (a + c)/(b + d)$, would have been a very helpful tool to already know, as we could have applied it directly. However, it is also something that we may derive to help us solve the problem.

Tools Used and Developed

- If $\frac{a}{b} = \frac{c}{d}$, then $\frac{a}{b} = \frac{a+c}{b+d}$.

5

Complex Numbers and Trigonometry

Problem 5.1

Calculate the product of tangent A for all angles of A from 1° to 89° in the triangle below.

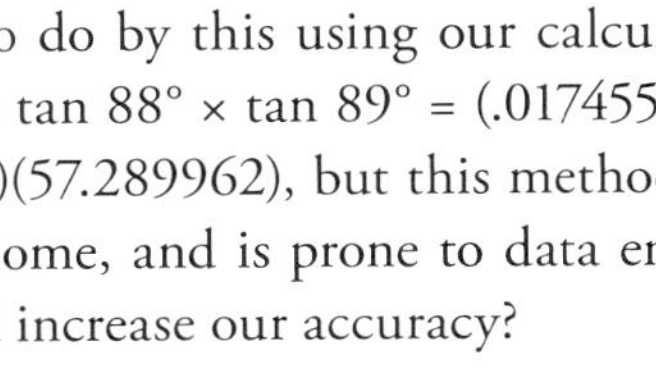

There is a direct way to do by this using our calculator, tan 1° × tan 2° × tan 3° × … × tan 87° × tan 88° × tan 89° = (.017455)(.034920)(.052408) … (19.081136)(28.636253)(57.289962), but this method would take us far too long, is rather cumbersome, and is prone to data entry error. How can we shorten the process and increase our accuracy?

Insight 1

Let's think about what we know regarding right triangles. We know

$$\sin A = \frac{a}{c}, \quad \cos A = \frac{b}{c}, \quad \tan A = \frac{a}{b}.$$

Knowing these relationships allows us to express tan A in terms of sin A and cos A:

$$\frac{\sin A}{\cos A} = \frac{\frac{a}{c}}{\frac{b}{c}} = \frac{a}{c} \times \frac{c}{b} = \frac{a}{b} = \tan A,$$

so we see that tan A = sin A/cos A. Now we can rewrite our equation as

$$\tan 1^\circ \times \tan 2^\circ \times \tan 3^\circ \times \ldots \times \tan 87^\circ \times \tan 88^\circ \times \tan 89^\circ$$

$$= \frac{\sin 1^\circ}{\cos 1^\circ} \times \frac{\sin 2^\circ}{\cos 2^\circ} \times \frac{\sin 3^\circ}{\cos 3^\circ} \times \ldots \times \frac{\sin 87^\circ}{\cos 87^\circ} \times \frac{\sin 88^\circ}{\cos 88^\circ} \times \frac{\sin 89^\circ}{\cos 89^\circ}.$$

In the middle of this stream, we have tan 45°, which equals 1, so we'll take that out of consideration, and just deal with the remaining expressions as above. We need one further fact, though, to make this pesky problem manageable.

We know that in a right triangle, the sum of angles A and B equals 90° and we know that sin A = cos (90° − A) = cos B, so now we can rewrite our equation one more time, expressing each sine in terms of cosine:

$$\frac{\sin 1^\circ}{\cos 1^\circ} \times \frac{\sin 2^\circ}{\cos 2^\circ} \times \frac{\sin 3^\circ}{\cos 3^\circ} \times \ldots \times \frac{\sin 87^\circ}{\cos 87^\circ} \times \frac{\sin 88^\circ}{\cos 88^\circ} \times \frac{\sin 89^\circ}{\cos 89^\circ}$$

$$= \frac{\cos 89^\circ}{\cos 1^\circ} \times \frac{\cos 88^\circ}{\cos 2^\circ} \times \frac{\cos 87^\circ}{\cos 3^\circ} \times \ldots \times \frac{\cos 3^\circ}{\cos 87^\circ} \times \frac{\cos 2^\circ}{\cos 88^\circ} \times \frac{\cos 1^\circ}{\cos 89^\circ}.$$

Now we see that the terms cancel each other out, and we arrive at our answer: 1.

Insight 2

Another approach is to look at the triangle and remind ourselves that tan A = a/b and that cot A = 1/tan A = b/a. But, we also see that tan B = b/a, so we have cot A = 1/tan A = tan B. Finally, B = (90° − A). Therefore, tan 89° = cot 1° = 1/tan 1°.

We can now write our original expression,

$$\tan 1^\circ \times \tan 2^\circ \times \tan 3^\circ \times \ldots \times \tan 87^\circ \times \tan 88^\circ \times \tan 89^\circ,$$

as

$$\tan 1^\circ \times \tan 2^\circ \times \tan 3^\circ \times \ldots \times \cot 3^\circ \times \cot 2^\circ \times \cot 1^\circ.$$

Recall that tan 45° in the middle equals 1, so we drop that, and we can pair the remaining 88 numbers in 44 pairs of tangent and cotangent. Since $\cot A = 1/\tan A$, we now see that our expression reduces to the product of a bunch of tangents and their reciprocals, which all cancel out to give us the simple answer of 1.

Amazing what a little insight does—the clever use of even simple trig identities can vastly simplify problems.

Tools Used and Developed

- In right triangle ABC, with right angle C, $\sin A = \cos(90^\circ - A) = \cos B$.
- In right triangle ABC, with right angle C, $\cot A = 1/\tan A = \tan B$.

Problem 5.2

Calculate

$$\sin^2 10^\circ + \sin^2 20^\circ + \sin^2 30^\circ + \sin^2 40^\circ + \sin^2 50^\circ + \sin^2 60^\circ + \sin^2 70^\circ + \sin^2 80^\circ + \sin^2 90^\circ.$$

This is the sort of problem that is really irritating—just busy work. I got the values of sine on my calculator, squared them, and summed them in memory. The values are shown in the table below, rounded to five decimal places. On my calculator, though, I worked with 32 significant figures.

	10°	*20°*	*30°*	*40°*	*50°*	*60°*	*70°*	*80°*	*90°*
sin	0.17365	0.34202	0.5	0.64279	0.76604	0.86603	0.93969	0.98481	1
$\sin^2$	0.03015	0.11698	0.25	0.41318	0.58682	0.75	0.88302	0.96985	1

To my surprise, when I retrieved the sum from memory, it was 5.

Now let's look at $\sin^2 1° + \sin^2 2° + \sin^2 3° + \ldots + \sin^2 88° + \sin^2 89° + \sin^2 90°$. There's no way we're going to sit and waste our time doing this! However, the fact that the value of our smaller problem was 5 strongly suggests that there may be a simpler way to get the answer. All we need is the right insight.

Insight

The insight comes from something we used for Problem 5.1. This is a great way to develop insights—apply something we've seen before in a new context. Perhaps we didn't see it the first time, but the second time around, we have a better chance of pulling out a tool that will help us find an insightful, even ingenious, solution to a problem.

We have already seen that $\sin A = \cos (90° - A)$ from examining the definitions of sine and cosine in a right triangle. Therefore, we can play around with substituting for some or all of the sine terms in our problem. In addition, the definitions of sine and cosine from Problem 5.1, combined with the Pythagorean theorem that $a^2 + b^2 = c^2$ immediately yield $\sin^2 A + \cos^2 A = 1$, which is the most basic trigonometric identity.

We see, for example, that $\sin 1° = \cos (90° - 1°) = \cos 89°$, so we can also say that $\sin^2 1° = \cos^2 89°$. Now we can start to substitute for some of the $\sin^2$ terms:

$$\sin^2 1° + \sin^2 2° + \sin^2 3° + \ldots + \sin^2 88° + \sin^2 89° + \sin^2 90°$$

can become

$$\cos^2 89° + \cos^2 88° + \ldots + \cos^2 46° + \sin^2 45° + \sin^2 46° + \ldots + \sin^2 88° + \sin^2 89° + \sin^2 90°.$$

We also see that we can to pair up the $\sin^2$ and $\cos^2$ terms, starting with $(\sin^2 46° + \cos^2 46°)$ and going through $(\sin^2 89° + \cos^2 89°)$, and be left with $\sin^2 45°$ and $\sin^2 90°$ as unpaired terms. So we now have

$$(\sin^2 46° + \cos^2 46°) + (\sin^2 47° + \cos^2 47°) + \ldots + (\sin^2 89° + \cos^2 89°) + \sin^2 45° + \sin^2 90°.$$

There are 44 sets of parentheses, each of the form $\sin^2 A + \cos^2 A$, which we know equals 1. Also, since $\sin 90° = 1$, then $\sin^2 90° = 1$. Finally, since $\sin 45° = \sqrt{2}/2$, then $\sin^2 45° = 1/2$. Therefore, the total sum is 45.5.

Tools Used and Developed

- Pairing to simplify calculations.
- $\sin^2 A + \cos^2 A = 1$.

Problem 5.3

Find the sum of the roots of the equation $\tan^2 x - 9\tan x + 1 = 0$ where $0 \le x \le \pi$ radians.

We recognize that this is a quadratic equation, and by letting $y = \tan x$, we can rewrite the equation as: $y^2 - 9y + 1 = 0$. By using the quadratic formula, we get

$$y = \tan x = \frac{9 \pm \sqrt{81-4}}{2} = \frac{9 \pm \sqrt{77}}{2}.$$

This looks rather messy, but we can plug it into our calculator and get (rounding off) the two values

$$x = \tan^{-1}(8.88748),$$

$$x = \tan^{-1}(0.112518).$$

Now, we use the arctan function on the calculator to get $x \approx 6.4198°$ and $x \approx 83.5802°$. Adding these up, we get a sum of about 90°.

However, we must also remember a key fact: $\tan(180° + \theta) = \tan\theta$. Therefore, there are two additional values of x for which x is less than 360° that we would get by adding 180° to each of the values we got from solving our quadratic. So now we have four solutions:

$$x \approx 6.4198°, \quad x \approx 83.5802°, \quad x \approx 186.4198°, \quad x \approx 263.5802°.$$

As we can see, the last two solutions are greater than $180° = \pi$ radians, and thus fall outside the range of x for our problem, where it was specified that $0 \le x \le \pi$ radians. Thus, adding up our allowed solutions, we get the sum as $90° = \pi/2$ radians.

If there is a simpler solution, we need an insight. We certainly can't waste this kind of time on one problem if, for example, we are taking the AHSME exam. By the time we solve it, the other kids will have already finished the test and gone home!

Insight 1

One good approach to try when we encounter tangent or cotangent in a problem is to reduce the expression to sine and cosine terms, as this often clarifies the approach we need to take. Rewriting the original problem, $y^2 - 9y + 1 = 0$, we get

$$\frac{\sin^2 x}{\cos^2 x} - 9\frac{\sin x}{\cos x} + 1 = 0.$$

Clearing the fractions by multiplying by $\cos^2 x$, we get

$$\sin^2 x - 9 \sin x \cos x + \cos^2 x = 0.$$

Wow! This is great. Right away, we see our most basic identity, $\sin^2 x + \cos^2 x = 1$, as part of our equation, which now reduces to $1 - 9 \sin x \cos x = 0$, or $\sin x \cos x = 1/9$.

We can simplify further. Going back to our toolbox of common trigonometric identities, we know that $\sin(A + B) = \sin A \cos B + \sin B \cos A$. If both A and B equal x, as they do here, then we get the convenient "double-angle" identity, $\sin(2x) = 2 \sin x \cos x$, or

$$\sin x \cos x = \frac{\sin(2x)}{2}.$$

Since our original equation has reduced to $\sin x \cos x = 1/9$, we now know that $\sin(2x)/2 = 1/9$, or $\sin(2x) = 2/9$. This gives us that $2x = \sin^{-1}(2/9)$. Before we waste our time actually figuring out $\sin^{-1}(2/9)$, we need to remember that since there is a first-quadrant angle, $\theta = \sin^{-1}(2/9)$, there will also be a second-quadrant angle $(\pi - \theta)$, whose sine also equals 2/9. Thus, we know that $2x$ can have two valid values, $2x = \sin^{-1}(2/9)$ and $2x = \pi - \sin^{-1}(2/9)$. From these equations, we see that there are two values of x that satisfy our equation,

$$x = \frac{1}{2}\sin^{-1}\frac{2}{9} \text{ and } x = \frac{\pi}{2} - \frac{1}{2}\sin^{-1}\frac{2}{9}.$$

Adding these up, we see that the $\sin^{-1}(2/9)$ portion cancels out, and we are left with the sum of the roots $x + x = 2x = \pi/2$ radians. Very clean and simple, with no messy inverse trig calculations.

Once again, as a mental check, we note that the other angle values that satisfy the requirement would be angles such as $2x = 2\pi + \sin^{-1}(2/9)$, which would give us values of x that are greater than π, falling outside our specified range.

Insight 2

There is an even easier solution, which I think is more elegant. Our original equation, $\tan^2 x - 9\tan x + 1 = 0$, is just a quadratic equation in terms of $\tan x$, which we rewrote as $y^2 - 9y + 1 = 0$, by letting $y = \tan x$. By Vieta's formula, which we looked at in Problem 4.2, we know that the product of the roots of this quadratic equals 1. We can see this easily by supposing that the roots are r and s. Then, we know that we can rewrite the quadratic in terms of its roots; i.e., $y^2 - 9y + 1 = (y - r)(y - s)$. If this product is expanded, we get that $rs = 1$.

This is huge! This tells us that the roots of the quadratic are reciprocals. Thus, if there is a value $x = \theta$ such that $\tan\theta$ is one of the roots of the quadratic, we know that $1/\tan\theta$ is also a root of the quadratic. Another name for $1/\tan\theta$ is $\cot\theta$, so we know that if $\tan\theta$ is one of the roots of the quadratic, then so is $\cot\theta$.

For our particular quadratic, we see that the solutions are

$$\tan x = \frac{9 \pm \sqrt{77}}{2},$$

which tells us that both values of $\tan x$ are positive, meaning that the angles x are either in the first or third quadrant. We ignore the third quadrant angles here since we know that for this problem we are interested only in values of x such that $0 \le x \le \pi$. Thus, focusing on the first-quadrant solutions (which must be present since all positive numbers have an inverse tangent in the first quadrant), we see that if $\tan\theta$ is a root, then so is $\cot\theta$. However, we know from right triangles that

$$\cot\theta = \tan\left(\frac{\pi}{2} - \theta\right).$$

Thus, if $x = \theta$ is a first-quadrant solution, then so is $x = (\pi/2) - \theta$. Adding these two solutions, we get that the sum is $\pi/2$, as we had obtained before. Not only did we not need to do any messy inverse trig calculations, we also didn't have to bother with potentially confusing double-angle identities.

Moreover, this solution lets us see something rather dramatic—the value of θ essentially doesn't matter at all (as long as θ is a first-quadrant angle) because it cancels out. Thus, we can confidently say that there was nothing special about our quadratic $\tan^2 x - 9\tan x + 1 = 0$. We would have gotten the same answer of $\pi/2$ for the sum of the values of x of any quadratic of the form $\tan^2 x - b\tan x + 1 = 0$, where b is some positive constant, since the roots of that quadratic would be

$$\tan x = \frac{b \pm \sqrt{b^2 - 4}}{2},$$

which are both positive (i.e, would have first-quadrant solutions).

Don't believe it? Let's do one the hard way. If $b = 10$, we have $\tan x = (10 \pm \sqrt{96})/2$, or $\tan x = 8.89898$ and $\tan x = 0.101021$. If we take inverse tangents of these values, we get $x \approx 84.23152°$ and $x \approx 5.76848°$, which sum to 90°. Pretty neat, huh?

Tools Used and Developed

- Expressing $\tan x$ in terms of $\sin x$ and $\cos x$.
- The double-angle identity, $\sin(2x) = 2\sin x \cos x$.
- $\sin\theta = \sin(\pi - \theta)$.
- Vieta's formula: for a quadratic of the form $y^2 + by + 1 = 0$, the product of the roots is 1.
- $\cot\theta = \dfrac{1}{\tan\theta}$, and $\cot\theta = \tan\left(\dfrac{\pi}{2} - \theta\right)$.

Problem 5.4

Let $a = \pi/2{,}008$. Find the sum of

$$\sin a \cos a + \sin 2a \cos 4a + \sin 3a \cos 9a + \cdots + \sin 10a \cos 100a.$$

First, note the number 2,008 in this problem. It comes from the 2008 AIME exam, and these contest guys are always so clever in working the contest year into their problems. Clever or not, we immediately see how painful this is going to be. We start by writing out the sum and doing some simplifications along the way:

$$\sin\left(\frac{\pi}{2{,}008}\right)\cos\left(\frac{\pi}{2{,}008}\right)+\sin\left(\frac{\pi}{1{,}004}\right)\cos\left(\frac{\pi}{502}\right)$$

$$+\sin\left(\frac{3\pi}{2{,}008}\right)\cos\left(\frac{9\pi}{2{,}008}\right)+\cdots+\sin\left(\frac{5\pi}{1{,}004}\right)\cos\left(\frac{25\pi}{502}\right).$$

Using a calculator that is turned to radians, we very painfully go through the process of calculating the numbers:

(0.00156454)(0.999999) + (0.00312907)(0.99998) + … + (0.0156447) (0.987786).

After adding them up and rewriting the problem in summation notation, we get

$$\sum_{i=1}^{10}\sin\left(\frac{i\pi}{2008}\right)\cos\left(\frac{i^2\pi}{2008}\right)=0.0856255.$$

That was horrible, but at least I think it is correct.

Now let's find the sum for $i = 100$:

$$\sum_{i=1}^{100}\sin\left(\frac{i\pi}{2008}\right)\cos\left(\frac{i^2\pi}{2008}\right).$$

This seems absolutely insane—we cannot possibly do this without a major insight.

Insight

We notice that this problem is a sum of terms of the form $\sin A \cos B$. Each one of these terms, by itself, is messy, but perhaps there is some way to work with the terms to make their sum a bit easier. In trigonometry, terms of this form (i.e., a product of basic trigonometric functions) can be rewritten as a sum of trigonometric functions. This is done by using something called a "product-to-sum" identity. We can derive this identity somewhat easily by using a yet more basic identity known as an "angle sum" identity. We will not derive this basic identity here, and instead consider it just one of the tools we have picked up in our general study of math. Thus, this insight is really a type 2 insight, which means that we would need to already be familiar with some basic tools to be able to come up with it.

The angle sum identity we start with is:

$$\sin(A+B) = \sin A \cos B + \sin B \cos A. \tag{5.1}$$

And we use the facts that $\sin(-B) = -\sin B$ and $\cos(-B) = \cos B$ to derive the sister identity,

$$\sin(A-B) = \sin A \cos B - \sin B \cos A. \tag{5.2}$$

Now we can get to work. We see the term we are interested in, $\sin A \cos B$, in each of these identities. We can isolate this term by adding Equations (5.1) and (5.2) to get

$$\sin A \cos B = \frac{1}{2}\left[\sin(A+B) + \sin(A-B)\right].$$

This is the product-to-sum identity.

Let's look now at our problem in general:

$$\sin a \cos a + \sin 2a \cos 4a + \sin 3a \cos 9a + \cdots + \sin na \cos n^2 a.$$

We can use the product-to-sum identity on each of these terms:

$$\sin a \cos a = \frac{1}{2}\left[\sin(a+a) + \sin(a-a)\right] = \frac{1}{2}\left[\sin 2a + \sin 0\right],$$

$$\sin 2a \cos 4a = \frac{1}{2}\left[\sin(2a+4a) + \sin(2a-4a)\right] = \frac{1}{2}\left[\sin 6a + \sin(-2a)\right],$$

$$\sin 3a \cos 9a = \frac{1}{2}\left[\sin(3a+9a) + \sin(3a-9a)\right] = \frac{1}{2}\left[\sin 12a + \sin(-6a)\right],$$

$$\sin 4a \cos 16a = \frac{1}{2}\left[\sin(4a+16a) + \sin(4a-16a)\right] = \frac{1}{2}\left[\sin 20a + \sin(-12a)\right],$$

$$\vdots$$

$$\sin na \cos n^2 a = \frac{1}{2}\left[\sin(na+n^2 a) + \sin(na-n^2 a)\right]$$

$$= \frac{1}{2}\left[\sin(n(n+1)a) + \sin(-n(n-1)a)\right].$$

Recalling that sin $(-B)$ = $-$sin B, we can rewrite sin $(-2a)$ = $-$sin $2a$, sin $(-6a)$ = $-$sin $6a$, and so on. We can rewrite our terms as

$$\sin a \cos a = \frac{1}{2}\left[\sin(a+a) + \sin(a-a)\right] = \frac{1}{2}\left[\sin 2a + \sin 0\right],$$

$$\sin 2a \cos 4a = \frac{1}{2}\left[\sin(2a+4a) + \sin(2a-4a)\right] = \frac{1}{2}\left[\sin 6a - \sin 2a\right],$$

$$\sin 3a \cos 9a = \frac{1}{2}\left[\sin(3a+9a) + \sin(3a-9a)\right] = \frac{1}{2}\left[\sin 12a - \sin 6a\right],$$

$$\sin 4a \cos 16a = \frac{1}{2}\left[\sin(4a+16a) + \sin(4a-16a)\right] = \frac{1}{2}\left[\sin 20a - \sin 12a\right],$$

$$\vdots$$

$$\sin na \cos n^2 a = \frac{1}{2}\left[\sin(na+n^2a) + \sin(na-n^2a)\right]$$

$$= \frac{1}{2}\left[\sin(n(n+1)a) - \sin n(n-1)a\right].$$

We see that if we add up these terms, something amazing happens: we get telescoping. The sin $2a$ term in the first bracket cancels with the $-$sin $2a$ term in the second bracket, the sin $6a$ term in the second bracket cancels with the $-$sin $6a$ term in the third bracket, and so on. Remembering that sin $0°$ = 0, we see that after we sum all the terms, we are left with only the left term in the last bracket: (1/2)sin $n(n + 1)a$. Therefore, we can say that

$$\sum_{i=1}^{n} \sin(ia)\cos\left(i^2 a\right) = \frac{1}{2}\sin n(n+1)a.$$

That is an amazing simplification, and means that we have to evaluate only one trig function. Going back to our first problem, we have $a = \pi/2{,}008$ and $n = 10$. Therefore, we claim that the first sum,

$$\sum_{i=1}^{10} \sin\left(\frac{i\pi}{2{,}008}\right)\cos\left(\frac{i^2\pi}{2{,}008}\right),$$

should equal $(1/2)\sin(110\pi/2{,}008)$. When we plug this into the calculator, we get 0.0856255, just as we did the painstaking way.

Now for our much longer summation, where $n = 100$, we are quite confident that the answer will be $(1/2)\sin(10{,}100\pi/2{,}008)$, which equals −0.0468672. Much shorter this way, isn't it?

Tools Used and Developed

- Angle sum identities:

  $\sin(A+B) = \sin A \cos B + \sin B \cos A$ and $\sin(A-B) = \sin A \cos B - \sin B \cos A$.

- Product-to-sum identity:

$$\sin A \cos B = \frac{1}{2}\left[\sin(A+B) + \sin(A-B)\right].$$

Problem 5.5

Evaluate

$$\frac{\cos 1^\circ + \cos 2^\circ + \cos 3^\circ + \cdots + \cos 43^\circ + \cos 44^\circ}{\sin 1^\circ + \sin 2^\circ + \sin 3^\circ + \cdots + \sin 43^\circ + \sin 44^\circ}.$$

This tricky problem, which comes from the AIME, could really bog us down. There is no mystery about what needs to be done here. We just have to do a lot of careful calculator work, getting and adding a lot of values, storing the results for the numerator and the denominator, and then dividing. This is very painful and time consuming—just what we have come to expect. I actually did this, and I got the fraction

$$\frac{\cos 1^\circ + \cos 2^\circ + \cos 3^\circ + \cdots + \cos 43^\circ + \cos 44^\circ}{\sin 1^\circ + \sin 2^\circ + \sin 3^\circ + \cdots + \sin 43^\circ + \sin 44^\circ} \approx \frac{39.660}{16.423} \approx 2.415.$$

After all that work, we are forced to ask: "Is there an insight that young Gauss could have used to get the answer more efficiently?"

Insight

Here, again, we are dealing with a type 2 insight—that is, the type that comes to us when we already have some familiarity with a particular tool, and then the insight tells us how to use that tool to our advantage. If we don't even know that the tool exists, we'll have a really hard time coming up with the needed insight. In Problem 5.4, for example, we made use of the idea that there are sum-to-product and product-to-sum identities in trigonometry.

Here, we see that both the numerator and the denominator of the problem are sums of trig functions, so we can start to think about using the sum-to-product identities. For the numerator, we are adding cosine terms, so let's derive the sum-to-product identity for cosines. We'll take as a given that we have previously learned the identities for the cosine of sums and differences:

$$\cos(A+B)=\cos A\cos B-\sin A\sin B, \tag{5.3}$$

$$\cos(A-B)=\cos A\cos B+\sin A\sin B. \tag{5.4}$$

If we add Equations (5.3) and (5.4) together, we get

$$\cos(A+B)+\cos(A-B)=2\cos A\cos B. \tag{5.5}$$

Now comes a bit of a trick. Define $A = (x + y)/2$, $B = (x - y)/2$. By doing this, we see that $A + B = x$ and $A - B = y$, and we can now rewrite Equation (5.5) as

$$\cos x+\cos y=2\cos\left(\frac{x+y}{2}\right)\cos\left(\frac{x-y}{2}\right).$$

This is known as the sum-to-product identity for cosines.

The questions now become: How do we apply this? Is there a convenient way to pair off the terms to simplify the problem? Right off the bat, we see that the terms can be combined into pairs that sum to 45° (e.g., 44° + 1°, 43° + 2°, and so on). We don't quite know where we're going yet, but because 45° is a familiar angle, we follow our nose and see where this approach might lead. We start paring the terms in the numerator and summing as follows:

$$\cos 44^\circ + \cos 1^\circ = 2\cos\left(\frac{44^\circ + 1^\circ}{2}\right)\cos\left(\frac{44^\circ - 1^\circ}{2}\right) = 2\cos\left(\frac{45^\circ}{2}\right)\cos\left(\frac{43^\circ}{2}\right),$$

$$\cos 43^\circ + \cos 2^\circ = 2\cos\left(\frac{43^\circ + 2^\circ}{2}\right)\cos\left(\frac{43^\circ - 2^\circ}{2}\right) = 2\cos\left(\frac{45^\circ}{2}\right)\cos\left(\frac{41^\circ}{2}\right),$$

$$\cos 42^\circ + \cos 3^\circ = 2\cos\left(\frac{42^\circ + 3^\circ}{2}\right)\cos\left(\frac{42^\circ - 3^\circ}{2}\right) = 2\cos\left(\frac{45^\circ}{2}\right)\cos\left(\frac{39^\circ}{2}\right),$$

$$\vdots$$

$$\cos 23^\circ + \cos 22^\circ = 2\cos\left(\frac{23^\circ + 22^\circ}{2}\right)\cos\left(\frac{23^\circ - 22^\circ}{2}\right) = 2\cos\left(\frac{45^\circ}{2}\right)\cos\left(\frac{1^\circ}{2}\right).$$

We're not quite sure whether we've made things better because there are still a lot of messy terms. We know how to handle the cos(45°/2) part with something called the half-angle identity, but the rest of the terms look rather daunting. However, we still need to apply the sum-to-product identity to the denominator. Once again, starting from our basic trig identities for the sine of angle sums, we derive the sum-to-product identity for sines:

$$\sin(A + B) = \sin A \cos B + \cos A \sin B,$$

$$\sin(A - B) = \sin A \cos B - \cos A \sin B.$$

Proceeding exactly as before, we get that

$$\sin x + \sin y = 2\sin\left(\frac{x+y}{2}\right)\cos\left(\frac{x-y}{2}\right).$$

Now we apply this to the sine terms in the same way as we did for the cosines, pairing them up to sum to 45°:

$$\sin 44^\circ + \sin 1^\circ = 2\sin\left(\frac{44^\circ + 1^\circ}{2}\right)\cos\left(\frac{44^\circ - 1^\circ}{2}\right) = 2\sin\left(\frac{45^\circ}{2}\right)\cos\left(\frac{43^\circ}{2}\right),$$

$$\sin 43^\circ + \sin 2^\circ = 2\sin\left(\frac{43^\circ + 2^\circ}{2}\right)\cos\left(\frac{43^\circ - 2^\circ}{2}\right) = 2\sin\left(\frac{45^\circ}{2}\right)\cos\left(\frac{41^\circ}{2}\right),$$

$$\sin 42^\circ + \sin 3^\circ = 2\sin\left(\frac{42^\circ + 3^\circ}{2}\right)\cos\left(\frac{42^\circ - 3^\circ}{2}\right) = 2\sin\left(\frac{45^\circ}{2}\right)\cos\left(\frac{39^\circ}{2}\right),$$

$$\vdots$$

$$\sin 23^\circ + \sin 22^\circ = 2\sin\left(\frac{23^\circ + 22^\circ}{2}\right)\cos\left(\frac{23^\circ - 22^\circ}{2}\right) = 2\sin\left(\frac{45^\circ}{2}\right)\cos\left(\frac{1^\circ}{2}\right).$$

We can now factor cos(45°/2) out of the numerator and sin(45°/2) out of the denominator and write our original fraction as

$$\frac{\cos 1^\circ + \cos 2^\circ + \cos 3^\circ + \cdots + \cos 43^\circ + \cos 44^\circ}{\sin 1^\circ + \sin 2^\circ + \sin 3^\circ + \cdots + \sin 43^\circ + \sin 44^\circ}$$

$$= \frac{2\cos\left(\frac{45^\circ}{2}\right)\left[\cos\left(\frac{43^\circ}{2}\right) + \cos\cdots\left(\frac{41^\circ}{2}\right) + \cos\left(\frac{39^\circ}{2}\right) + \cdots + \cos\left(\frac{1^\circ}{2}\right)\right]}{2\sin\left(\frac{45^\circ}{2}\right)\left[\cos\left(\frac{43^\circ}{2}\right) + \cos\left(\frac{41^\circ}{2}\right) + \cos\left(\frac{39^\circ}{2}\right) + \cdots + \cos\cdots\left(\frac{1^\circ}{2}\right)\right]}.$$

In an almost miraculous fashion, we see an amazing cancellation of the troublesome cosine terms in the brackets, and all of our problems go away. We are left with

$$\frac{\cos 1^\circ + \cos 2^\circ + \cos 3^\circ + \cdots + \cos 43^\circ + \cos 44^\circ}{\sin 1^\circ + \sin 2^\circ + \sin 3^\circ + \cdots + \sin 43^\circ + \sin 44^\circ} = \frac{2\cos\left(\frac{45^\circ}{2}\right)}{2\sin\left(\frac{45^\circ}{2}\right)}$$

$$= \cot\left(\frac{45^\circ}{2}\right) \approx 2.414$$

(using a calculator).

To be a little fancier, we can simplify the expression to a closed form by using something called the half-angle identities. Let us start again with our most basic angle sum identities:

$$\sin(A + B) = \sin A \cos B + \cos A \sin B,$$

$$\cos(A + B) = \cos A \cos B - \sin A \sin B.$$

If we let A and B both equal y, we have

$$\sin(2y) = 2\sin y \cos y,$$

$$\cos(2y) = \cos^2 y - \sin^2 y = 2\cos^2 y - 1.$$

If we let $y = x/2$, and rearrange the terms, we get

$$\sin x = 2\sin\frac{x}{2}\cos\frac{x}{2}, \tag{5.6}$$

$$1 + \cos x = 2\cos^2\frac{x}{2}. \tag{5.7}$$

Dividing Equation (5.7) by Equation (5.6), we then get

$$\frac{2\cos^2\dfrac{x}{2}}{2\sin\dfrac{x}{2}\cos\dfrac{x}{2}} = \cot\frac{x}{2} = \frac{1+\cos x}{\sin x}.$$

Thus, if we let $x = 45°$, we get

$$\cot\frac{45°}{2} = \frac{1+\cos 45°}{\sin 45°}.$$

Now we don't even need a calculator. We have

$$\cot\frac{45°}{2} = \frac{1+\dfrac{\sqrt{2}}{2}}{\dfrac{\sqrt{2}}{2}} = \frac{2+\sqrt{2}}{\sqrt{2}} = \frac{2\sqrt{2}+2}{2} = \sqrt{2}+1.$$

Thus, we have arrived at a simple, elegant closed-form solution:

$$\frac{\cos 1° + \cos 2° + \cos 3° + \cdots + \cos 43° + \cos 44°}{\sin 1° + \sin 2° + \sin 3° + \cdots + \sin 43° + \sin 44°} = \sqrt{2}+1 \approx 2.414.$$

We see that this agrees numerically with the solution we derived the hard way. This way may have seemed like a lot of work also, but that is because we spent time deriving almost all of our identities along the way. When these identities are tools at our fingertips, however, the solution actually goes rather quickly. In any case, this was certainly much more fun than the brute force approach.

Tools Used and Developed

- The most basic trig identities regarding the cosine and sine of sums of angles:

$$\cos(A+B) = \cos A \cos B - \sin A \sin B$$

$$\cos(A-B) = \cos A \cos B + \sin A \sin B$$

$$\sin(A+B) = \sin A \cos B + \cos A \sin B,$$

$$\sin(A-B) = \sin A \cos B - \cos A \sin B.$$

- The sum-to-product identities:

$$\cos x + \cos y = 2\cos\left(\frac{x+y}{2}\right)\cos\left(\frac{x-y}{2}\right),$$

$$\sin x + \sin y = 2\sin\left(\frac{x+y}{2}\right)\cos\left(\frac{x-y}{2}\right).$$

- The half-angle identity for cotangent:

$$\cot\frac{x}{2} = \frac{1+\cos x}{\sin x}$$

- As a bonus, since $\tan(x/2) = 1/\cot(x/2)$, the half-angle identity for tangent,

$$\tan\frac{x}{2} = \frac{\sin x}{1+\cos x}.$$

Problem 5.6

Evaluate $(1 + i)^{20}$.

We can start off by calculating the first few powers to see whether a pattern emerges.

$$
\begin{aligned}
(1+i)^1 &= 1+i, \\
(1+i)^2 &= 1+2i+i^2 = 2i, \\
(1+i)^3 &= 2i(1+i) = -2+2i, \\
(1+i)^4 &= (-2+2i)(1+i) = -2-2i+2i-2 = -4, \\
(1+i)^5 &= -4(1+i) \\
&\vdots \\
(1+i)^8 &= (1+i)^4 \times (1+i)^4 = -4 \times -4 = 16.
\end{aligned}
$$

We can see that each fourth power of $(1 + i)$ is an integer power of -4. Specifically, the pattern is that $(1 + i)^{4k} = (-4)^k$. Since $(1 + i)^{20} = (1 + i)^{4\times5}$, we know that our expression is equal to $(-4)^5 = -1{,}024$. Therefore, $(1 + i)^{20} = -1{,}024$.

It was fortunate for us that the expression turned out to have an easy pattern. But what if we're not so lucky next time, and we are given an expression such as $(3+i\sqrt{3})$, and asked to raise that to the 20th, or even higher, power? We can look at the first few powers and try to discern some kind of pattern:

$$
\begin{aligned}
(3+i\sqrt{3})^1 &= 3+i\sqrt{3}, \\
(3+i\sqrt{3})^2 &= 9+6i\sqrt{3}-3 = 6+6i\sqrt{3}, \\
(3+i\sqrt{3})^3 &= (6+6i\sqrt{3})(3+i\sqrt{3}) = 18+6i\sqrt{3}+18i\sqrt{3}-18 = 24i\sqrt{3}, \\
(3+i\sqrt{3})^4 &= +24i\sqrt{3} \times (3+i\sqrt{3}) = 72i\sqrt{3}-72.
\end{aligned}
$$

There seems to be no easy pattern emerging, so we look for another way to tackle this problem. We need an insight.

Insight

This particular insight requires a more advanced tool pertaining to complex numbers, namely, the polar form of complex numbers. Similar to Cartesian rectangular and polar coordinates, the polar form of complex numbers is simply another way of expressing a complex number given in standard form. Just as polar coordinates in the Cartesian plane have a magnitude and direction, polar coordinates in the complex plane deal with the distance from the complex number to the origin as well as the direction in which that distance is traveled. Given any complex number, $a + bi$, the distance from this point to the origin, called the magnitude of the complex number, is $r = \sqrt{a^2 + b^2}$, and the direction angle is $\theta = \tan^{-1}(b/a)$. Therefore, the real part of the complex number is $r \cos \theta$, and the imaginary part is $r \sin \theta$. The imaginary number can thus be written as $r \cos \theta + ir \sin \theta = r(\cos \theta + i \sin \theta)$. The expression $\cos \theta + i \sin \theta$ is often abbreviated as $\text{cis}\ \theta$. Therefore, any complex number can be written in the form $r \text{ cis } \theta$.

Now that we have figured out another way of expressing a complex number, we need to figure out how to deal with raising the polar form to powers. To do this, we need one more slightly advanced tool, known as Euler's theorem, which states that $\text{cis}\ \theta$ is in fact equal to $e^{i\theta}$, where e is the famous mathematical constant equal to about 2.718. Now we are ready to start working.

If we want to raise a complex number $\cos \theta + i \sin \theta$ to a power, we can raise $e^{i\theta}$ to the same power, and the two expressions should be equal to each other: $(\cos \theta + i \sin \theta)^n = (e^{i\theta})^n = e^{in\theta}$. By Euler's theorem, $e^{in\theta} = \cos n\theta + i \sin n\theta$. Hence, $(\cos \theta + i \sin \theta)^n = \cos n\theta + i \sin n\theta$ (this is known as De Moivre's theorem). Aha! Now we have an easy way of raising any complex number to any power and calculating the result. We can now start to apply this to our problem. Our first step is to convert our complex number $(3 + i\sqrt{3})$ into polar form.

To find the magnitude of $(3 + i\sqrt{3})$, we take $\sqrt{a^2 + b^2}$, where a is 3 and b is $\sqrt{3}$. Therefore, the magnitude is $\sqrt{9+3} = \sqrt{12} = 2\sqrt{3}$. Now that we have the magnitude, we need to find the direction. To do this, we use $\tan^{-1}(b/a)$, which, when we substitute in our values for a and b, becomes $\tan^{-1}(3/\sqrt{3}) = \tan^{-1}\sqrt{3} = 60°$, or $\pi/3$ radians. Now that we have the magnitude and direction, we can write the complex number in polar form: $2\sqrt{3}\,\text{cis}(\pi/3)$. Using De Moivre's theorem, we have an easy method of calculating this complex number raised to the 20th power:

$$\left(2\sqrt{3}\operatorname{cis}\left(\frac{\pi}{3}\right)\right)^{20} = \left(2\sqrt{3}\right)^{20} \times \operatorname{cis}\left(\frac{20\pi}{3}\right) = 2^{20} \times 3^{10} \times \operatorname{cis}\left(\frac{2\pi}{3}\right)$$

$$= 61{,}917{,}364{,}220\left(\cos\frac{2\pi}{3} + i\sin\frac{2\pi}{3}\right)$$

$$= 61{,}917{,}364{,}220\left(-\frac{1}{2} + \frac{\sqrt{3}}{2}i\right)$$

$$= -30{,}958{,}682{,}110 + 30{,}958{,}682{,}110i\sqrt{3}.$$

That was much easier than having to multiply each power separately until we get to 20.

Problem 5.7

Evaluate

$$\sin\left(\frac{\pi}{5}\right) + \sin\left(\frac{2\pi}{5}\right) + \sin\left(\frac{3\pi}{5}\right) + \sin\left(\frac{4\pi}{5}\right).$$

No problem, we think. First, we convert to degrees, as it's a bit easier to punch into a calculator, and we get $\sin(36^\circ) + \sin(72^\circ) + \sin(108^\circ) + \sin(144^\circ)$. Plugging into the calculator and rounding off, we get $0.58779 + 0.95106 + 0.95106 + 0.58779$, which is approximately 3.0777.

Of course, we could have made this a bit simpler by realizing that $\sin(\pi - \theta) = \sin(\theta)$ so $\sin(36^\circ) = \sin(144^\circ)$ and $\sin(72^\circ) = \sin(108^\circ)$. Thus, we really needed to figure out only $\sin(36^\circ)$ and $\sin(72^\circ)$. In any case, this was an easy enough problem to whip out.

So we try to find

$$\sin\left(\frac{\pi}{100}\right) + \sin\left(\frac{2\pi}{100}\right) + \sin\left(\frac{3\pi}{100}\right) + \cdots + \sin\left(\frac{99\pi}{100}\right).$$

If we use the same approach for this longer problem, it would take forever—we would need to mess with fractional degrees, round off, figure out fifty different values of sin, double them, and add them up. We need an insight.

Insight

This is the kind of insight that is based on knowing a specific mathematical tool or theorem—what we have called a type 2 insight. If we don't have this tool (i.e., if we are not familiar with this tool or theorem), then it would be very difficult for us to have the required insight.

But we do know how to do sums of arithmetic series and geometric series in a nice, abbreviated fashion. Too bad this series is neither. However, if we remember that sin and cos functions are related to the exponential representation of imaginary numbers by Euler's theorem, and that this leads directly to De Moivre's theorem, we can crack this problem by converting it into a geometric series summation.

As usual, we recognize that there is likely nothing special about the numbers in the denominator, so we'll tackle the general problem of

$$\sin\left(\frac{\pi}{n}\right)+\sin\left(\frac{2\pi}{n}\right)+\sin\left(\frac{3\pi}{n}\right)+\cdots+\sin\left(\frac{(n-1)\pi}{n}\right).$$

Let us remind ourselves of Euler's theorem, which is this absolutely amazing statement that $e^{i\theta} = \cos\theta + i\sin\theta$. Given this, we see that:

$$(\cos\theta + i\sin\theta)^n = (e^{i\theta})^n = e^{in\theta} = \cos n\theta + i\sin n\theta,$$

which is De Moivre's theorem. Therefore, if we let $z = \cos\theta + i\sin\theta$, then we see that $\sin\theta$ is just the imaginary part of z, which we can call $\mathrm{Im}(z)$, and that $\sin k\theta$ is just $\mathrm{Im}(z^k)$. Moreover, we see that $z^{-1} = \cos(-\theta) + i\sin(-\theta) = \cos\theta - i\sin\theta$, since $\cos(-\theta) = \cos\theta$ and $\sin(-\theta) = -\sin\theta$. Given these facts, we can also see that

$$\cos\theta = \frac{z+z^{-1}}{2} \quad \text{and} \quad \sin\theta = \frac{z-z^{-1}}{2i}.$$

These are just general mathematical facts, but extremely important ones.

For our problem, let us say

$$\theta = \frac{\pi}{2n}, \quad so \quad z = \cos\left(\frac{\pi}{2n}\right) + i\,\sin\left(\frac{\pi}{2n}\right).$$

We see that $\sin(k\pi/n) = \mathrm{Im}(z^{2k})$. Also, we see that $z^{2n} = \cos\pi + i\sin\pi = -1$. We recall that if we sum a bunch of complex numbers, the sum of the imaginary parts is just the imaginary part of the sum, so we can rewrite our desired sum as:

$$\sum_{k=1}^{n-1} \sin\left(\frac{k\pi}{n}\right) = \mathrm{Im}\sum_{k=1}^{n-1} z^{2k}.$$

Voila! We've transformed our problem into a geometric series with common ratio z^2 and first term z^2. We know how to sum this series (so we won't review that again here!). We can now say:

$$\mathrm{Im}\sum_{k=1}^{n-1} z^{2k} = \mathrm{Im}\frac{z^2 - z^{2n}}{1 - z^2}.$$

We have to keep remembering that we are focusing on the imaginary part of this expression, because this is what represents the sin terms. Now we remember our nifty little fact that $z^{2n} = -1$, and we have

$$\mathrm{Im}\sum_{k=1}^{n-1} z^{2k} = \mathrm{Im}\frac{z^2 + 1}{1 - z^2}.$$

Let's multiply the numerator and denominator by z^{-1} to get

$$\mathrm{Im}\sum_{k=1}^{n-1} z^{2k} = \mathrm{Im}\frac{z^2 + 1}{1 - z^2} = \mathrm{Im}\frac{z + z^{-1}}{z^{-1} - z}.$$

Recalling that $\cos\theta = (z + z^{-1})/2$ and $\sin\theta = (z - z^{-1})/(2i)$, and that we have defined $\theta = \pi/2n$, we can rewrite our result as

$$\operatorname{Im}\frac{2\cos\left(\dfrac{\pi}{2n}\right)}{-2i\sin\left(\dfrac{\pi}{2n}\right)}.$$

Multiplying the numerator and denominator by i, and canceling the 2s, we see that the result is $\operatorname{Im}[i\cot(\pi/2n)]$. Since this is a purely imaginary number, the imaginary part of it is the whole thing, and we have arrived at a very beautiful and compact result:

$$\sin\left(\frac{\pi}{n}\right)+\sin\left(\frac{2\pi}{n}\right)+\sin\left(\frac{3\pi}{n}\right)+\cdots+\sin\left(\frac{(n-1)\pi}{n}\right)=\cot\left(\frac{\pi}{2n}\right).$$

If we go to our first simple problem, $\sin(\pi/5) + \sin(2\pi/5) + \sin(3\pi/5) + \sin(4\pi/5)$, we see that this should be just $\cot(\pi/10) = \cot(18°)$. This gives us about 3.0777, the same result as the brute force method.

Now we can tackle our harder problem,

$$\sin\left(\frac{\pi}{100}\right)+\sin\left(\frac{2\pi}{100}\right)+\sin\left(\frac{3\pi}{100}\right)+\cdots+\sin\left(\frac{99\pi}{100}\right).$$

We know this is just $\cot(\pi/200)$, which is about 63.6567.

This saved a ton of work!!

Tools Used and Developed

- Euler's theorem, $e^{i\theta} = \cos\theta + i\sin\theta$.
- De Moivre's theorem, $(\cos\theta + i\sin\theta)^n = \cos n\theta + i\sin n\theta$.
- If we represent a complex number as $z = \cos\theta + i\sin\theta$, then

$$\cos\theta=\frac{z+z^{-1}}{2}\quad\text{and}\quad\sin\theta=\frac{z-z^{-1}}{2i}.$$

- Summing a geometric series.

Problem 5.8

Evaluate cos 5° + cos 10° + cos 15° + cos 20° + cos 25° + cos 30°.

We look at the problem, thinking about whether there are easy or obvious simplifications by using double-angle or half-angle identities, but nothing comes to mind. Anyway, the problem is easy enough to do by brute force, using our calculator (rounding to six decimal places):

0.996195 + 0.984808 + 0.965926 + 0.939693 + 0.906308 + 0.866025 = 5.658954.

It was a bit cumbersome rounding and adding these numbers, but not fatal.

Now we want to consider cos 1° + cos 2° + cos 3° + cos 4° + ⋯ + cos 50°. This method is just unacceptable now. It would really be a huge waste of time to do this by brute force. We need an insight.

Insight

Using what we learned from Problem 5.7, we start by generalizing this problem and thinking about an exponential representation of trig functions. Since there seems to be nothing special about the angle measures in terms of double- or half-angle identities, and we notice that the angles simply increase in regular increments, the most general version of the problem would be:

$$\cos\theta + \cos 2\theta + \cos 3\theta + \cos 4\theta + \ldots + \cos n\theta. \tag{5.8}$$

We remind ourselves of the basic facts that helped us crack Problem 5.7:

- Euler's theorem: $e^{i\theta} = \cos\theta + i\sin\theta$.
- De Moivre's theorem: $(\cos\theta + i\sin\theta)^n = (e^{i\theta})^n = e^{in\theta} = \cos n\theta + i\sin n\theta$.
- If we let $z = \cos\theta + i\sin\theta$, then

$$\cos\theta = \frac{z+z^{-1}}{2} \quad \text{and} \quad \sin\theta = \frac{z-z^{-1}}{2i}.$$

In Problem 5.7, we had great success working with something of the form $z = e^{i\theta/2}$, and we were able to turn the problem into a geometric series with common ratio z^2. So let's try that again. If we let $z = e^{i\theta/2}$, we see that

$\cos\theta = (z^2 + z^{-2})/2$, $\cos 2\theta = (z^4 + z^{-4})/2$, and so on, with $\cos n\theta = (z^{2n} + z^{-2n})/2$. Therefore, Equation (5.8) can be written as

$$\begin{aligned}&\cos\theta + \cos 2\theta + \cos 3\theta + \cos 4\theta + \cdots + \cos n\theta \\ &= \frac{1}{2}\left(z^{-2n} + z^{-(2n-2)} + \cdots + z^{(2n-2)} + z^{2n},\right)\end{aligned} \tag{5.9}$$

where we have arranged the terms in increasing powers.

There is something very important here that is easy to overlook. In the middle of the series in parentheses will be a term of the form z^0, as we progress from z^{-2n} to z^{2n}. This term, z^0, is equivalent to 1, and has not been accounted for in the left-hand cosine series of Equation (5.9). Therefore, to be correct, we have to add 1/2 to the left-hand side of Equation (5.9) to get

$$\begin{aligned}&\frac{1}{2} + \cos\theta + \cos 2\theta + \cos 3\theta + \cdots + \cos n\theta \\ &= \frac{1}{2}\left(z^{-2n} + z^{-(2n-2)} + \cdots + z^{(2n-2)} + z^{2n}\right).\end{aligned}$$

The series in parentheses is now a simple geometric series with a starting term of z^{-2n}, a final term of z^{2n}, and a common ratio of z^2.

If we sum this series and multiply the sum by 1/2, we get

$$\frac{1}{2}\left(z^{-2n} + z^{-(2n-2)} + \cdots + z^{(2n-2)} + z^{2n}\right) = \frac{z^{-2n} - z^{2n+2}}{2(1-z^2)} = \frac{z^{2n+2} - z^{-2n}}{2(z^2-1)}.$$

Here is where the magic of using the common ratio of z^2 becomes apparent. Let us multiply the numerator and denominator by z^{-1}, and we get:

$$\frac{1}{2}\left(z^{-2n} + z^{-(2n-2)} + \cdots + z^{(2n-2)} + z^{2n}\right) = \frac{z^{2n+1} - z^{-(2n+1)}}{2(z - z^{-1})}. \tag{5.10}$$

Recalling that we defined $z = e^{i\theta/2}$, we see that Equation (5.10) simplifies to

$$\frac{1}{2}\left(z^{-2n} + z^{-(2n-2)} + \cdots + z^{(2n-2)} + z^{2n}\right) = \frac{2i\sin\left(\frac{(2n+1)\theta}{2}\right)}{2\cdot 2i\sin\cdots\left(\frac{\theta}{2}\right)} = \frac{\sin\left(\frac{(2n+1)\theta}{2}\right)}{2\sin\left(\frac{\theta}{2}\right)}.$$

However, we have to recall that

$$\frac{1}{2}\left(z^{-2n}+z^{-(2n-2)}+\cdots+z^{(2n-2)}+z^{2n}\right)$$

$$=\frac{1}{2}+\cos\theta+\cos 2\theta+\cos 3\theta+\cdots+\cos n\theta,$$

so our series sum is

$$\cos\theta+\cos 2\theta+\cos 3\theta+\cdots+\cos n\theta=\frac{\sin\left(\frac{(2n+1)\theta}{2}\right)}{2\sin\left(\frac{\theta}{2}\right)}-\frac{1}{2}.$$

We have solved our problem. For the simple first problem, $\cos 5^\circ + \cos 10^\circ + \cos 15^\circ + \cos 20^\circ + \cos 25^\circ + \cos 30^\circ$, we have $\theta = 5^\circ$ and $n = 6$. Therefore, we expect to have

$$\cos 5^\circ+\cos 10^\circ+\cos 15^\circ+\cos 20^\circ+\cos 25^\circ+\cos 30^\circ=\frac{\sin(32.5^\circ)}{2\sin(2.5^\circ)}-\frac{1}{2}.$$

Indeed, when we punch this into our calculator, we get 5.658954, just as we did before.

Now comes the time to reap the benefits of our insight. We needed to sum $\cos 1^\circ + \cos 2^\circ + \cos 3^\circ + \cos 4^\circ + \cdots + \cos 50^\circ$. Here, $\theta = 1^\circ$, and $n = 50$. Thus, our sum should be

$$\frac{\sin(50.5^\circ)}{2\sin(0.5^\circ)}-\frac{1}{2},$$

which works out to 43.711393.

Tools Used and Developed

- Euler's theorem: $e^{i\theta} = \cos\theta + i\sin\theta$.
- De Moivre's theorem: $(\cos\theta + i\sin\theta)^n = (e^{i\theta})^n = e^{i\theta} = \cos n\theta + i\sin n\theta$.

- If we let $z = \cos\theta + i\sin\theta$, then we see that

$$\cos\theta = \frac{z + z^{-1}}{2} \quad \text{and} \quad \sin\theta = \frac{z - z^{-1}}{2i}.$$

- Turning a sum of a series of cosines into a geometric series by using the above tools. In particular, the bright idea of letting $z = e^{i\theta/2}$ so we could turn the problem into a geometric series with common ratio z^2. This made it simple to convert the result back to elementary trig functions.

6

Changing Perspective

Problem 6.1

What is the first time after 12 o'clock noon at which the hour and minute hands on a clock meet?

Our first realization is that the meeting will happen at a time that is a little after 1 o'clock.

Thinking about the problem algebraically, we figure it would best be solved by coming up with equations in terms of time for the positions of the hour and minute hands, and setting those equations equal to figure out at what precise time those positions are equal. Since we are dealing with a circular path, namely the clock, it would be best to think of the positions in terms of degrees.

We know that, for constant velocity, the position of an object at a certain time t is the starting position of the object plus the object's velocity multiplied by t. We know that the starting positions of the hands are at 12 o'clock, which is equivalent to 0°. We still, however, need to figure out the rates at which they travel.

The minute hand travels one whole revolution of the clock, or 360°, in 60 minutes. Therefore, it travels 6° every minute. Using the same logic, the hour hand travels 1/12 of a revolution, or 30° in 60 minutes, or half a degree every minute. Therefore, our position equations for each hand are

$$P_m = 0 + 6t = 6t,$$

$$P_h = 0 + \left(\frac{1}{2}\right)t = \left(\frac{1}{2}\right)t.$$

We are now in a good position to solve this problem.

Since we don't know how far either hand has traveled, we can just assign a variable to represent this. Let's let θ equal the number of degrees that the hour hand has traveled away from the 12 o'clock position. We know then that the minute hand has traveled that same θ, plus another 360° because it makes one full revolution around the clock before returning to the 12 o'clock position and starting on its way back to meet the hour hand. Therefore, the minute hand travels $\theta + 360$. Plugging these into our position equations, we get

$$P_m = \theta + 360 = 6t, \tag{6.1}$$

$$P_h = \theta = \left(\frac{1}{2}\right)t. \tag{6.2}$$

Equation (6.2) gives us an expression for θ purely in terms of t, which we can plug into Equation (6.1) to get

$$P_m = \theta + 360 = \left(\frac{1}{2}\right)t + 360 = 6t,$$

which simplifies to $(11/2)t = 360$, so $t = 720/11$ minutes $= 65\frac{5}{11}$ minutes. Therefore, the hands of the clock meet at about 27 seconds after 1:05 p.m.

Even though this is not a terribly grueling problem to solve using algebra, there are still easier and more innovative ways to solve it that put us in the mindset needed to come up with creative insights for future harder problems.

Insight

The following perspective change is actually quite a fun way to think about the problem. Instead of assuming the point of view of the observer of the clock who just watches two hands run around until they meet, and sets up equations describing their motions relative to himself, why don't we start off by pretending that we are the hour hand?

How is the problem different if we look at it from this perspective? If we are the hour hand, we don't know that we are moving; all we know is that after some period of time, the minute hand comes to visit us and then leaves us only to return a little while later. Let's begin by thinking about the first meeting. We know that it happens after a little over an hour. Once we meet, we can think of it as resetting the problem (i.e., the point at which we meet is the new 12 o'clock position at which we started). From this, we realize that the next meeting will take the same amount of time as the first, and that, therefore, the meetings occur at constant intervals. Now let's think about 12 o'clock midnight. At this point, it has been 12 hours, or 720 minutes, since the problem began. Counting all of the times we meet from noon to midnight, we see that there are 11 visits: one slightly after 1 p.m., one slightly after 2 p.m., etc. So we have 11 visits in 720 minutes, and since the meetings occur at constant rates, each one takes 720/11 minutes.

This solution required no equations and virtually no arithmetic. It was simply a creative change of perspective that allowed us to solve this problem.

Tools Used and Developed

- Change of perspective.

Problem 6.2

Let us say that a certain city holds a citywide tennis tournament. Because there are so many players, the tournament is run as follows: for the first round, the participants are randomly paired by a computer to play matches. If there is an odd number of players, one player, selected at random, gets a "bye," meaning that the player automatically goes on to the second round. After the first round is played, all the winners advance, and are once again randomly paired up to play the second round matches. Again, if there is an odd number, a player (at random) gets a bye into the third round, and so the process continues. The winners move forward, and randomly get paired up, until finally a city tennis champion is crowned. In our city, there were 2,011 entrants this year. How many matches were played until a champion was crowned?

To do this problem and not make mistakes, we have to be very systematic, proceeding step-by-step.

Step 1: The computer picks 1 person at random for a bye, and pairs the remaining 2,010 contestants into 1,005 matches. These will produce 1,005 winners who will move to round 2.
Step 2: The 1,006 people (1,005 winners from round 1 and the 1 person who got a bye) are paired into 503 matches for round 2. These produce 503 winners who move to round 3.
Step 3: Of the 503 winners, 1 person is selected to get a bye into round 4, and the remaining 502 are paired to play 251 matches. These produce 251 winners who move to round 4.
Step 4: The 251 winners and the 1 person who got a bye are now paired into 126 matches for round 4. These produce 126 winners.
Step 5: The 126 winners from round 4 are paired into 63 matches for round 5. These produce 63 winners who move to round 6.
Step 6: One person is selected randomly for a bye into round 7, and the remaining 62 are paired randomly into 31 matches. These produce 31 winners.
Step 7: The 31 winners from round 6 and the person who got a bye are paired to play 16 matches. These produce 16 winners.
Step 8: The 16 winners from round 7 are paired to play 8 matches, which produce 8 winners.
Step 9: The 8 winners are paired to play 4 matches, which produce 4 winners who move on.
Step 10: The 4 winners are paired to play 2 matches, which produce 2 winners.
Step 11: The last two players play 1 final match to crown a champion.

We have kept close track of all the matches, and underlined the number of matches played in each round. To find the total, we add them up:

Total Matches = 1,005 + 503 + 251 + 126 + 63 + 31 + 16 + 8 + 4 + 2 + 1 = 2,010 matches.

Most people, by the way, get lost somewhere along the way, and end up making a mistake in the math.

Insight

Most people will think the purpose of a match is to produce a winner who will go on to the next round, and they will follow the matches and the winners (as we did above). However, a clever change of perspective can greatly simplify the problem. Let us say that the purpose of a match is to produce a loser, who will be eliminated from the tournament. Each match has exactly one loser, so there is a one-to-one correspondence between the number of matches and the number of losers who are then eliminated. If we have 2,011 players, then to crown a champion, we will need 2,010 losers, and hence there will be 2,010 matches played. The problem is solved without any computation whatsoever, simply based on pure insight.

Tools Used and Developed

- Change of perspective.

Problem 6.3

The largest possible circle is inscribed into a square. In the circle is inscribed the largest possible square that will fit. Figure out the ratio of the areas of the two squares.

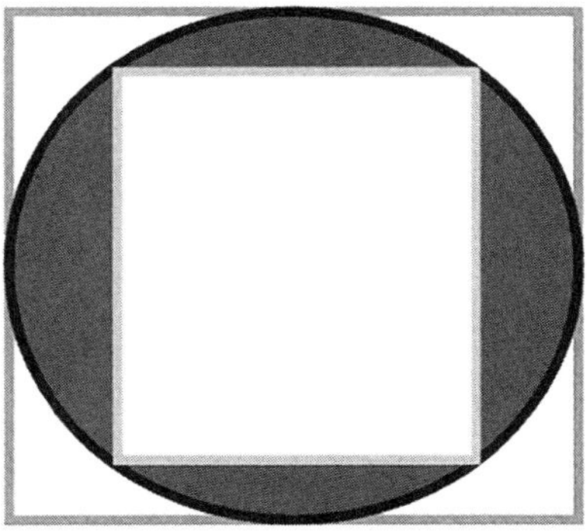

Let's start with what we know. If the side of the outer square measures S, we know that the area of the outer square is S^2. What else do we know?

Given that the circle is the largest possible circle that will fit into the outer square, we know that the circle's diameter is also S, so its radius has to be $S/2$. Knowing this, we can use the equation $A = \pi r^2$ to find the area of the circle,

$$A = \pi r^2 = \pi\left(\frac{S}{2}\right)^2,$$

and we can find the ratio of the area of the circle to the area of the larger square,

$$\frac{\text{circle}}{\text{outer square}} = \frac{\pi\left(\frac{S}{2}\right)^2}{S^2} = \frac{\pi}{4}.$$

Now let's take another look at our diagram and focus on the inner square with side s.

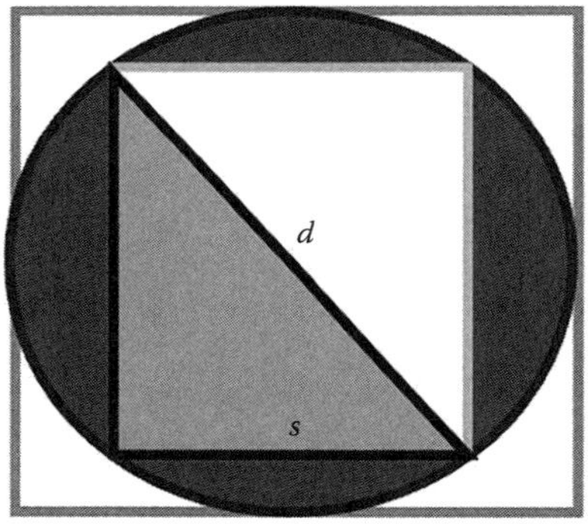

We know that the area of the inner square is s^2. Additionally, we also know that the diagonal d of the inner square is equal to the diameter of the circle. Hence, the radius of the circle is $d/2$, and the area of the inner circle is

$$\pi r^2 = \pi\left(\frac{d}{2}\right)^2 = \pi\frac{d^2}{4}.$$

Using the Pythagorean theorem, let's now express the area of the inner square in terms of d:

$$s^2 + s^2 = d^2$$

$$2s^2 = d^2$$

$$s^2 = \frac{d^2}{2}.$$

Now let's look at the ratio of the inner square to the circle:

$$\frac{\text{inner square}}{\text{circle}} = \frac{\frac{d^2}{2}}{\pi\frac{d^2}{4}} = \frac{d^2}{2} \times \frac{4}{\pi d^2} = \frac{2}{\pi}.$$

Using what we have calculated thus far, we can find the ratio of the inner square to the outer square:

$$\frac{\text{circle}}{\text{outer square}} \times \frac{\text{inner square}}{\text{circle}} = \frac{\text{inner square}}{\text{outer square}}, \quad \text{or} \quad \frac{\pi}{4} \times \frac{2}{\pi} = \frac{1}{2}.$$

So we can see that the area of the inner square is one-half the area of the outer square.

That certainly was doable, but somewhat tedious. Let's see if changing our vantage point might provide us with an insight that would make things a little easier.

Insight

We start with our original diagram:

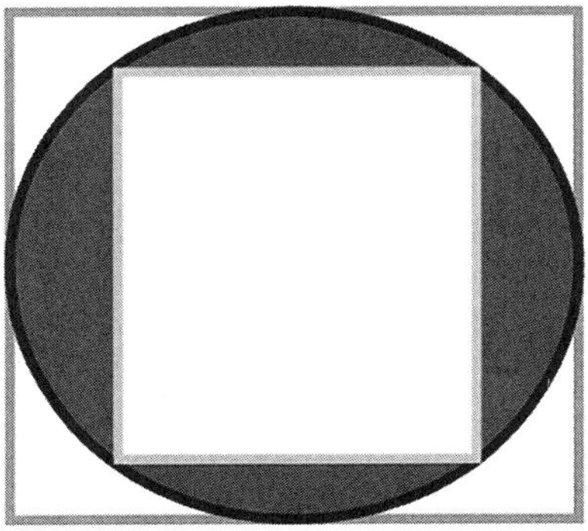

What if we rotate the inner square inside the circle so that it looks more like a diamond and we divide the outer square into four smaller squares?

By doing this, we have also divided the inner square (the diamond) into four congruent triangles. We can see that each of these four congruent triangles is half the area of the smaller square in which it lies. Thus, we can readily see that the area of the inner square is half the area of the outer square without needing to do any tedious calculations.

Tools Used and Developed

- Change of perspective.

Problem 6.4

Imagine that we have 15 pebbles in a pile. We divide the pile into two smaller piles of any size, multiply the number of pebbles in each of the two smaller piles, and store this result in a running tally. We continue this process of dividing each of the available piles into two smaller piles, multiplying the number of pebbles in the divided piles, and adding the result to the tally until all of the piles have been divided up such that there is now only one pebble in each pile, and no further divisions are possible. What are all the possible sums of products in the tally?

This is quite an unusual problem. It seems that there are myriad possibilites, since there are so many ways to divide the piles at each step. The best thing to do in these situations is to start by just exploring the problem. Therefore, we can start by doing just one of the many possible scenarios, as diagramed below.

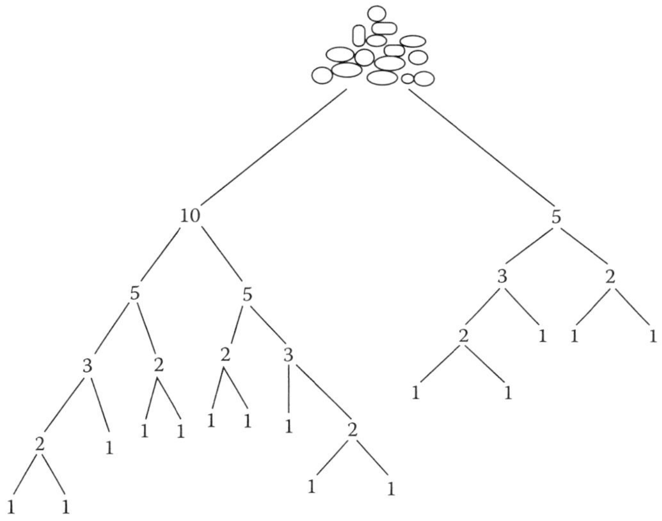

We start by dividing the 15 pebbles into two piles of 10 and 5 (first product is 50). We can see that the right and left branches now become separate, not affecting each other. Therefore, we can tally up the left branch first, and then tally up the right, rather than trying to follow both at the same time. On the left, we divide the 10-pebble pile into 5 and 5 (product 25), divide each of the piles of 5 into 3 and 2 (two products of 6), divide each of the piles of 3 into 2 and 1 (two products of 2), and divide each of the four piles of 2 that we now have into 1 and 1 (four products of 1). Therefore, for the left-hand branch, we have 25 + (2 × 6) + (2 × 2) + (4 × 1) = 45.

Doing the same approach for the right-hand branch, we get 6 + 2 + (2 × 1) = 10.

Now we add both of these to 50, which we got at the first branch point, and we get: 50 + 45 + 10 = 105.

Now what? This was just for an initial division of 10 and 5. We know that we have 7 ways to have made just that initial division of 15 pebbles [(1, 14), (2, 13), (3, 12), …., (7, 8)]. Of course, we see that it doesn't matter whether we look at a division of (5, 10), or (10, 5)—the results would be entirely symmetric if the left-hand branch had 5 pebbles and the right-branch had 10 in the diagram above.

The problem just mushrooms after the first branch, with so many different possibilites. It would be extremely time consuming to tally up all of the possibilities. And that's for just 15 pebbles—imagine if we want to do the same thing for 100 pebbles in a pile. Obviously, we need an insight.

Insight 1

Now that we have gone through one calculation for 15 pebbles with an initial division into piles of 10 and 5 (and gotten an answer of 105 in our tally of products), we want to explore what other answers are possible.

An approach that we may not think of right away is to always just take away 1 pebble. This is simple, allowed, and actually, quite brilliant. We start off dividing the 15 pebbles into two piles: 14 and 1. The product here is 14. There is nothing we can do with the pile that has just 1 pebble in it, so we go to the pile that has 14. Now we divide it into two piles: 13 and 1 (the product here is 13), and we keep going in this way, always taking just 1 pebble off. We can see that this will give us the following sum of products:

$$(14 \times 1) + (13 \times 1) + (12 \times 1) + \ldots + (1 \times 1) = 14 + 13 + 12 + \ldots + 1.$$

In other words, this is just the sum of the first 14 integers. This takes us back to Problem 1.1, and we know how to solve this in general:

$$S = \sum_{i=1}^{n} i = n(n+1)/2.$$

Here, then, the sum would be 14(15)/2 = 105. This is quite a surprise! This is the same answer that we got before, although we divided the piles in entirely different ways. Could it be that this is always the answer? This finding spurs us onto another insight as to how to prove this conjecture.

Insight 2

From our work so far, we notice that for n pebbles, the answer seems to be

$$\sum_{i=1}^{n-1} i = (n-1)n/2.$$

Imagine that all of the pebbles in a pile are connected by thin strings. In other words, each pebble is connected to every other pebble by a thin string, just like vertex points in a polygon are all connected to each other by edges or diagonals. Let's say we start with pile A, and divide it into two smaller piles, B and C. That means we need to cut all of the strings that connect the pebbles in pile B to those in pile C. This is just the product of the number of pebbles in pile B and the number of pebbles in pile C, because each pebble in pile B is connected to all of the pebbles in pile C. This product is exactly what we want to tally and sum.

As we keep subdividing the piles, we keep cutting strings. To get to the point where we have only one pebble in each pile, we have cut all of the strings. Thus, the total in our tally is the same as the total number of strings. In a starting pile of n pebbles, this is just

$$\sum_{i=1}^{n-1} i = (n-1)n/2.$$

Thus, we can see that for a pile of 15 pebbles, the answer is always 105, no matter how we subdivide the piles along the way. For a pile of 100 pebbles, then, the answer is just

$$\sum_{i=1}^{99} i = (99)100/2 = 4{,}950.$$

Both of our insights resulted from looking at the problem a bit differently, and one insight led to another. First, we thought of a limiting case of removing just one pebble at a time, and saw that this gave a very simple answer. Then, we decided to pretend that the pebbles were connected by strings, and related that problem of vertex points of a polygon connected by edges and diagonals, and saw that the products we needed to tally would sum up to the total number of strings, which is the same as the total number of connections between points in an equivalent polygon.

Tools Used and Developed

- Looking at things differently.

Problem 6.5

Two ferryboats travel at constant speeds back and forth across a river, turning around in zero time as soon as they reach the river bank. They leave at the same time from opposite shores, and meet 700 feet from one bank. They continue on their ways to the banks, turn around when they get there, and meet a second time 400 feet from the opposite bank. How wide is the river?

This sounds like a velocity or rates algebra problem. First, we have to translate the word problem into equations. Nothing helps us do that better than a simple diagram.

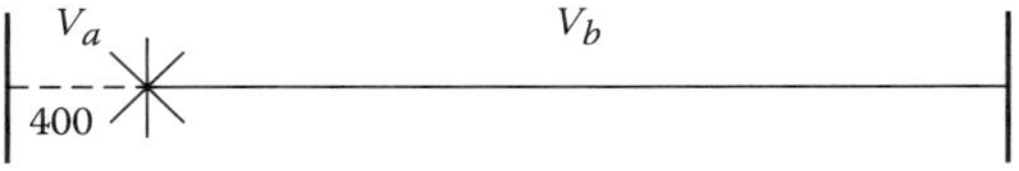

Let us say that boat A, traveling with speed V_a, leaves the right shore at the same time as boat B, traveling with speed V_b, leaves the left shore. Boat A is represented by the dashed arrow, and boat B is represented by the solid arrow. After a time t_1, they meet for the first time 700 feet from the right shore. They each continue on their way to the opposite bank, turn around and meet for the second time 400 feet from the left shore, as shown above. Assume that time t_2 elapses between the first and second meetings.

Now we can set up some equations. First, we know that $V_a t_1 = 700$ and that $V_b t_1 = W - 700$, where W is the width of the river. In other words, in time interval t_1, boat A has traveled 700 feet. We don't know how far boat B has traveled in this time interval, but we know that together, they have traveled the width of the river, W. Therefore, we can also say that

$$V_a t_1 + V_b t_1 = W. \qquad (6.1)$$

Let's look at time interval t_2, between the first and second meetings. Boat A continues traveling to the left, reaches the left bank, then turns around and travels 400 feet before meeting boat B. Therefore, it covers the distance traveled by boat B in the first time interval, plus an additional 400 feet. Thus, we can say that

$$V_a t_2 = V_b t_1 + 400. \tag{6.2}$$

By the same reasoning, we can also say that

$$V_b t_2 = V_a t_1 + (W - 400). \tag{6.3}$$

By substituting $V_b t_1 = W - 700$ into Equation (6.2), we get

$$V_a t_2 = W - 300, \tag{6.4}$$

and substituting $V_a t_1 = 700$ into Equation (6.3), we get

$$V_b t_2 = W + 300. \tag{6.5}$$

Dividing Equation (6.4) by $V_a t_1 = 700$, we get $t_2/t_1 = (W - 300)/700$, leading to

$$t_2 = \left(\frac{W - 300}{700}\right) t_1.$$

Dividing Equation (6.4) by Equation (6.5), we get $V_a/V_b = (W - 300)/(W + 300)$, so

$$V_b = \frac{W + 300}{W - 300} V_a. \tag{6.6}$$

Using Equation (6.1) and substituting Equation (6.6) for V_b and $V_a t_1 = 700$, we have $700 + [(W + 300)/(W - 300)]V_a t_1 = W$ and using $V_a t_1 = 700$ again, we now have

$$700+\left(\frac{W+300}{W-300}\right)700=W$$

$$700\left[\left(\frac{W+300}{W-300}\right)+1\right]=W$$

$$700\left(\frac{2W}{W-300}\right)=W$$

$$\frac{1400W}{W-300}=W,$$

which leads to $W = 1{,}700$ ft.

That was quite a bit of algebra and manipulation, but we got our answer. Is there an easier way to do this problem? If there is, it will need an insight.

Insight

Let us change our perspective entirely away from algebra, which caught us up because we realized that this was a rate-distance problem. Instead of looking at the time periods t_1 and t_2 separately, let's look at the distance that both boats travel together during the whole time.

We know that boat A ends up traveling the whole width of the river, and turning around and traveling another 400 feet, or $W + 400$ feet. Meanwhile, boat B travels the whole width of the river, and turns around and travels almost the whole way again, except for 400 feet. Therefore, it travels $2W - 400$ feet. Thus, together, both boats travel $3W$ feet. We know that when the boats first met, A had traveled 700 feet, and together the boats had traveled the width of the river, W. Since the boats move at constant speed, and since in total they traveled $3W$, we then know that boat A, in total, traveled 2,100 feet (since it traveled 700 feet when the two boats together had traveled W). Since A's total travel was 400 feet more than the width of the river, then $W = 1{,}700$. This was done very easily, without any algebra or serious calculations at all!

Tools Used and Developed

- Change of perspective. (Specifically, we looked at both boats together instead of trying to analyze equations for each boat separately.)

Problem 6.6

Now that we are comfortable with infinite series, let's try this famous problem, often referred to as "the fly and the trains" problem:

Two trains are 20 miles apart traveling toward each other on the same track. Each train travels at 10 miles per hour. A fly leaves the first train heading toward the second train. The fly flies at a constant speed of 15 miles per hour. When the fly reaches the second train, it turns and heads back toward the first train (assume that the turnaround takes no time at all). When the fly reaches the first train, it turns around again toward the second train. The fly keeps flying back and forth between the two trains until they collide, and the fly is crushed. What is the total distance the fly flies before it gets crushed?

Insight 1

We see that the fly keeps zigzagging between the trains, flying smaller and smaller distances each time. To get the total distance the fly travels, we need to sum all of these distances. Although we know the fly eventually gets crushed, we envision this mathematically as an infinite series of smaller distances. We feel good about this, since we now have some experience with series. We just hope it's a series we know how to sum!

To get a feel for what is happening, let's draw a very simple line diagram. We'll solve the problem in a completely general way. That's a bit harder, but I think we're up to it.

```
|------------------------------------d0------------------------------------|
|---Dt-->                                                           <--Dt---|
|---------|-------------------------------------------------------|--------|
|-----------------------------------Df---------------------------->|
```

The trains are initially separated by a distance d_0. The fly travels at speed v_f and the trains travel at speed v_t.

On the first leg, the fly leaves the left-hand train and flies to the right a distance D_f, landing on the nose of the train coming toward it, which has traveled to the left a distance D_t. This takes a certain amount of time, which we'll call t. We don't know the value of t, but we can say the following (referring to the diagram):

$$D_f = v_f t, \tag{6.7}$$

$$D_t = v_t t, \tag{6.8}$$

$$d_0 = D_f + D_t. \tag{6.9}$$

Dividing Equation (6.7) by Equation (6.8), we can say that $D_f = (v_f/v_t)D_t$, or alternatively, $D_t = (v_t/v_f)D_f$. Substituting for D_t in Equation (6.9), we can then write $D_f + (v_t/v_f)D_f = d_0$, or

$$D_f = \frac{d_0}{1+\dfrac{v_t}{v_f}} = \left(\frac{v_f}{v_f + v_t}\right)d_0. \tag{6.10}$$

Thus, the fly flies a fraction $v_f/(v_f + v_t)$ of the distance d_0 separating the trains on its first leg.

By a similar manipulation, we can establish that

$$D_t = \left(\frac{v_t}{v_f + v_t}\right)d_0. \tag{6.11}$$

As the fly turns around, it finds itself in a very similar situation, flying from the nose of one train to the other, except that the distance between the two trains is now smaller. If each train has moved a distance D_t during time t, the new separation is now $d_0 - 2D_t$. However, $d_0 = D_f + D_t$, so the new separation is now $D_f - D_t$.

Substituting in for D_f and D_t, we can write the new distance as

$$\left(\frac{v_f - v_t}{v_f + v_t}\right)d_0.$$

This is the distance facing the fly as it begins the second leg of the journey, with one of the trains coming toward it. Since the speeds have not changed, we know that the fly will cover a fixed fraction $v_f/(v_f + v_t)$ of this distance, on the second leg. This process will continue, with the distance between the trains shrinking by a factor of $(v_f - v_t)/(v_f + v_t)$ with each leg.

Therefore, we can set up the total distance flown by the fly as the sum of an infinite series

$$D=\left(\frac{v_f}{v_f+v_t}\right)d_0\left[1+\left(\frac{v_f-v_t}{v_f+v_t}\right)+\left(\frac{v_f-v_t}{v_f+v_t}\right)^2+\left(\frac{v_f-v_t}{v_f+v_t}\right)^3+\ldots\right]. \quad (6.12)$$

Let $(v_f - v_t)/(v_f + v_t) = r$. Thus, in the brackets, we have an infinite series of the form $S = 1 + r + r^2 + r^3 + \ldots$. This is a geometric series with common ratio r, and we know that it converges, because r is less than 1.

As we have covered before, then, $S = 1/(1 - r)$, or

$$S=\frac{1}{1-\left(\frac{v_f-v_t}{v_f+v_t}\right)}.$$

This simplifies to

$$S=\frac{1}{\left(\frac{(v_f+v_t)-(v_f-v_t)}{v_f+v_t}\right)}=\frac{v_f+v_t}{2v_t}.$$

Thus, substituting this value for the series in Equation (6.12), we get

$$D=\left(\frac{v_f}{v_f+v_t}\right)d_0\left(\frac{v_f+v_t}{2v_t}\right),$$

or finally

$$D=d_0\left(\frac{v_f}{2v_t}\right).$$

We now have a nice general solution. Plugging in our numbers, $d_0 = 20$, $v_f = 15$, and $v_t = 10$, we get $D = 15$ miles.

Insight 2

We were able to sum the infinite series and get a nice answer. However, there is a much slicker way to get the answer immediately. Since the trains are traveling toward each other, each with a speed of 10 mph, they are closing the distance at a speed of 20 mph (we can think of one train standing still and the other traveling toward it at 20 mph). Since the distance between the trains is 20 miles, it will take them 1 hour to meet. Since the fly flies at a speed of 15 mph, it will travel 15 miles in that hour!

In fact, the general solution can be obtained just as easily. If the trains are traveling toward each other with speed v_t, then it takes them time $t = d_0/(2v_t)$ to meet. Since the fly flies with speed v_f, the total distance it travels in this time is $v_f t = v_f[d_0/(2v_t)]$, the same answer we got for Insight 1.

Let's end this problem with a very amusing story about the mathematician John von Neumann. He was given this problem by a friend who knew the slick solution, and who wanted to see von Neumann struggle with the infinite series. To his friend's chagrin, von Neumann gave the right answer almost immediately.

"Interesting," said his friend, "most people try to sum the infinite series."

"What do you mean?" von Neumann replied. "Is there another way?"

Tools Used and Developed

- $S = 1/(1 - r)$, the sum of a geometric series $S = 1 + r + r^2 + r^3 + \ldots$.
- Flexibility and creativity to be able to see insightful solutions.

Problem 6.7

Two circles B and C, of radii 15 and 20, respectively, intersect as shown in the diagram below such that $\angle BAC = 90°$. What is the difference between the nonoverlapping areas of the two circles?

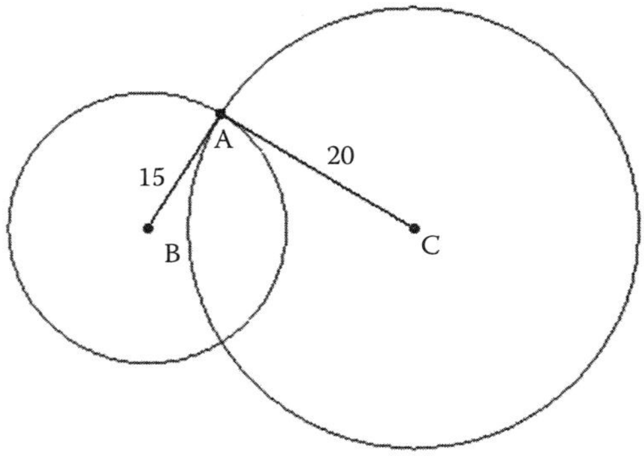

We can approach this problem by trying to find the areas of the nonoverlapping part of B and the nonoverlapping part of C and subtract them from each other. To do this, we need to find the area of the overlapping region, which is a portion of each circle. Our first instinct when trying to find the area of a portion of a circle might be to create a sector, so do that in circle B, as shown below. Of course, an analogous sector can be drawn in circle C (which will be done later).

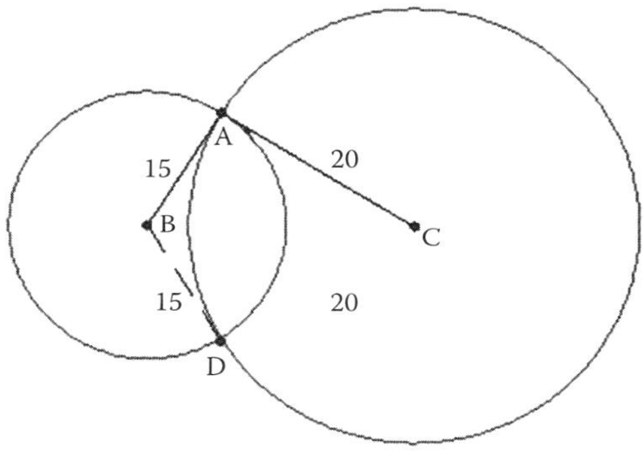

We see from the above diagram that the sector can be thought of as the combination of a triangle, triangle ABD in the diagram below, and a portion of the intersection between the two circles.

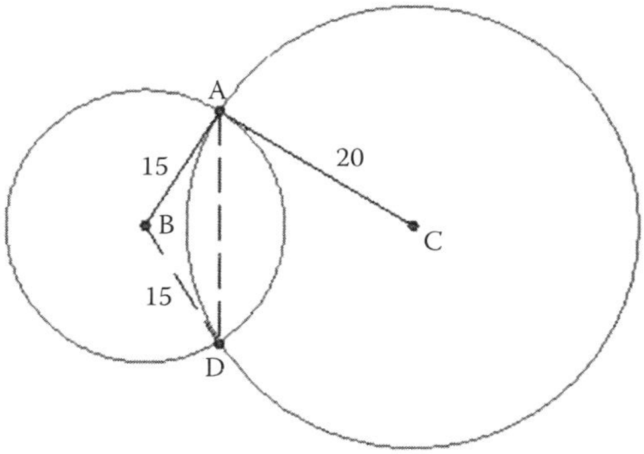

In looking at our diagram now, we have an epiphany! If we were to find the area of our sector on the circle with radius 15, and subtract away the area of triangle ABD, we would be left with the area of a portion of the intersecting section, shown in the diagram below.

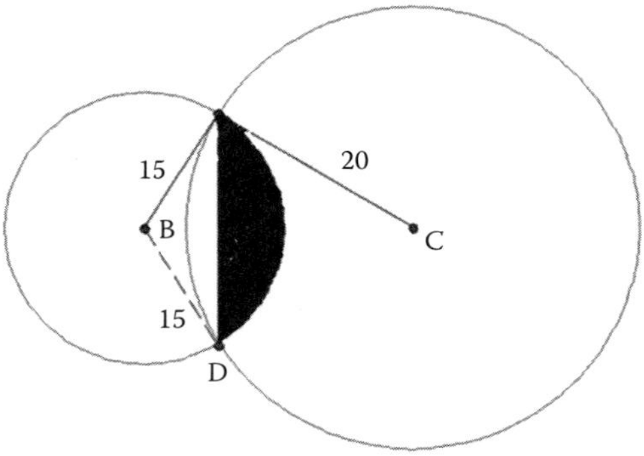

Now we are getting close. We set out to find the area of the intersecting portion of the circles and we have now found a part of it. We can also see that if we were to set up a sector on the circle with radius 20, we can find the area of the other part of the intersection between the circles.

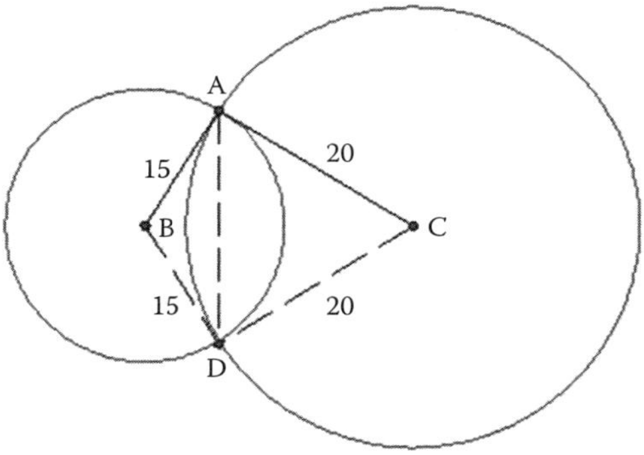

So if we were to find the area of the sector of the circle C and subtract away the area of triangle ACD, we would have the area of the remaining part of the intersecting region.

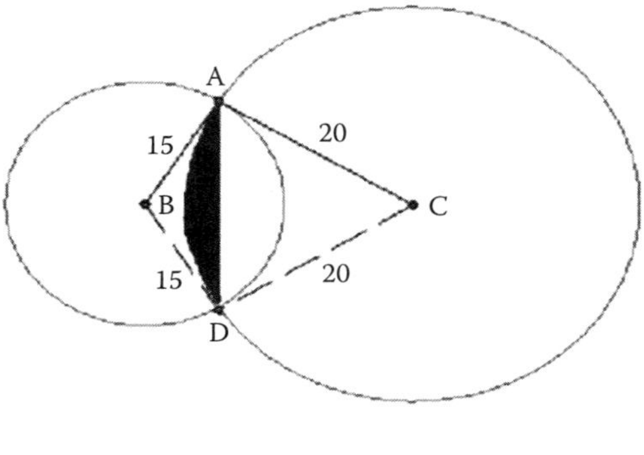

Thus, we have created the following diagram.

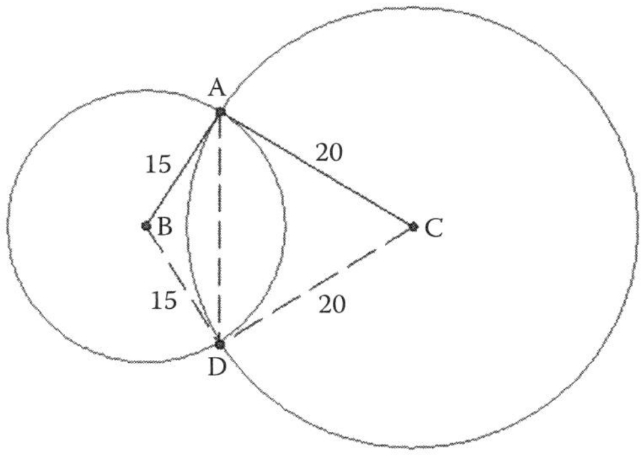

Given this diagram, we see that in order to find the area of the intersecting region, we must find the following:

> (Sector of Circle B – Area of Triangle ABD) + (Sector of Circle C – Area of Triangle ACD).

To calculate all of these areas, we need to find a few key pieces of information, namely ∠ACD, ∠ABD, and the length of AD, keeping in mind that ∠CAB is a right angle.

Using this information, we can create a right triangle, as shown below.

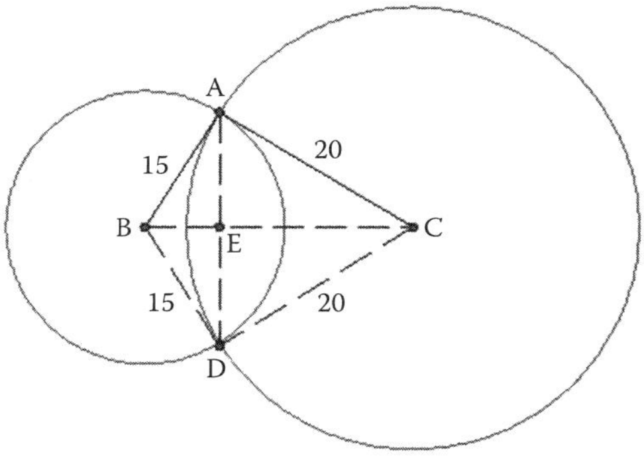

Using our knowledge of the Pythagorean theorem, we can figure out the length of segment BC.

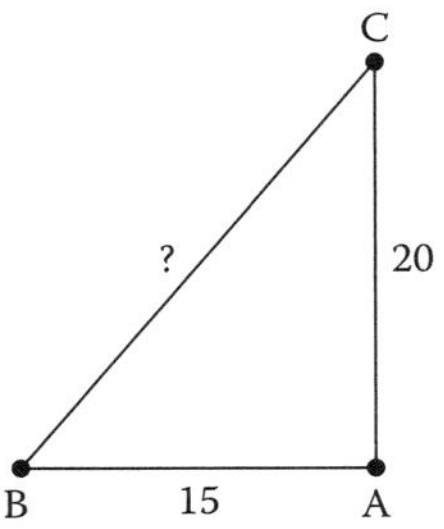

$$c^2 = a^2 + b^2 = 15^2 + 20^2 = 625 = c^2, \text{ so } c = 5.$$

Thus, we know that our right triangle looks like this.

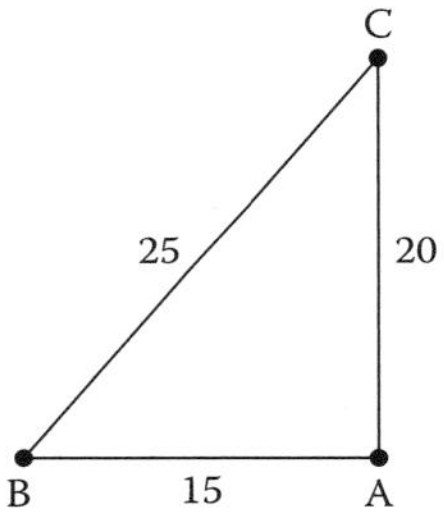

We can now label our circle diagram with this new information.

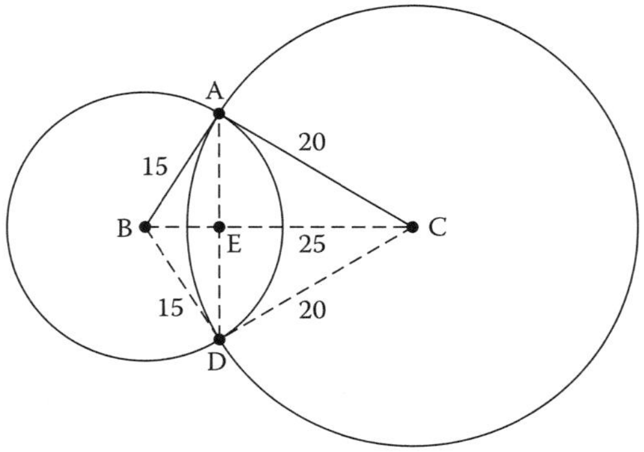

And we can now see that triangles ABC and BCD are 3-4-5 right triangles. The angles of a 3-4-5 right triangle are 37°-53°-90°, so we know that ∠ACB and ∠BCD are 37° and ∠ABC and ∠CBD are 53°. This means that ∠ABD is 106° and ∠ACD is 74°, and we can now find the areas of the two sectors.

$$\text{Sector ABD} = \frac{106}{360} \times \pi(15)^2 = 208.1,$$

$$\text{Sector ACD} = \frac{74}{360} \times \pi(20)^2 = 258.3.$$

Now we just need to find the areas of triangles ABD and ACD. To do this, we first find out the length of segment AD. Using the law of cosines, which states that if a triangle has side lengths X, Y, and Z, and corresponding angles x, y, and z, then $X^2 = Y^2 + Z^2 - 2(Y)(Z)\cos x$, we can thus calculate the length of segment AD:

$$\text{AD}^2 = 15^2 + 15^2 - 2(15)(15)\cos 106 = 701.6$$

$$\text{AD} = \sqrt{701.6} = 26.5.$$

Now we can use what is known as Heron's formula to find the area of triangles ABD and ACD. Heron's formula states that if the semiperimeter of a triangle is $s(= (a + b + c)/2)$ and the triangle has side lengths a, b, and c, the area of the triangle is $\sqrt{s(s-a)(s-b)(s-c)}$. So

$$\text{Area ABD} = \sqrt{28.25(28.25-15)(28.25-15)(28.25-26.5)} = 93.6$$

and

$$\text{Area ACD} = \sqrt{33.25(33.25-20)(33.25-20)(33.25-26.5)} = 198.5.$$

At last, we can find the area of the intersecting region.

$$\begin{aligned}\text{Area of Intersection} &= (\text{Sector of Circle B} - \text{Area of Triangle ABD}) \\ &\quad + (\text{Sector of Circle C} - \text{Area of Triangle ACD}) \\ &= (208.3 - 93.6) + (258.1 - 198.5) = 174.3.\end{aligned}$$

Using this value, we can now find the area of the nonintersecting regions of the two circles. For Circle B, the area of the nonintersecting region is $15^2\pi - 174.3 = 532.6$. For Circle C, the area of the nonintersecting region is $20^2\pi - 174.3 = 1{,}082.3$, so the difference between the nonoverlapping regions of the two circles is $1{,}082.3 - 532.6 = 549.7$.

By using this method, we were able to solve the problem. However, it was quite a tedious task. There is a much simpler and more elegant way to approach the problem that saves much time and effort.

Insight

Let's go back to the first diagram and say that the area of circle B is b. Likewise, the area of circle C is c. Let us also label the intersecting region x. We thus have the following diagram:

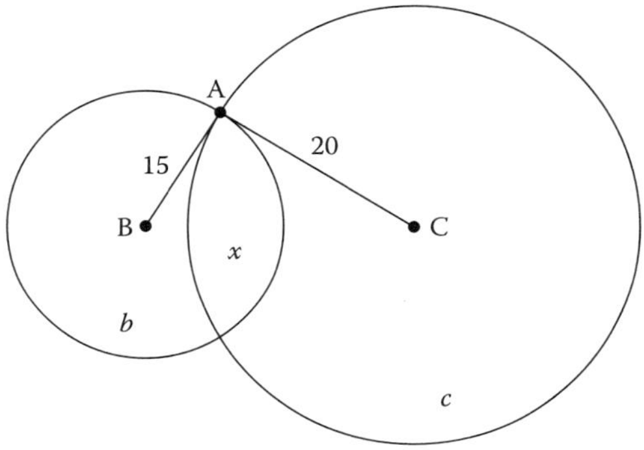

The expression for the nonoverlapping region of circle B is now $b - x$ and the area of the nonoverlapping region of circle C is $c - x$. Then the area for the difference between the nonintersecting areas of the triangles is simply

$$(c-x)-(b-x)=c-b=20^2\pi-15^2\pi=549.8. \qquad (6.13)$$

Notice that we obtain the same answer using both methods, except for a small rounding error. Notice also that the area of the intersecting region, x, canceled out in Equation (6.13). The area of the intersecting region doesn't matter. All of the effort expended in the first solution was unnecessary.

Tools Used and Developed

- Law of cosines: if a triangle has side lengths X, Y, and Z, and corresponding angles x, y, and z, then $X^2 = Y^2 + Z^2 - 2(Y)(Z)\cos x$.
- Heron's formula: if the semiperimeter of a triangle is s $(= (a + b + c)/2)$ and the triangle has side lengths a, b, and c, the area of the triangle is $\sqrt{s(s-a)(s-b)(s-c)}$.

- Keeping an eye on the big picture. We were never actually asked to find the area of the intersecting region. By keeping that in mind, we could solve the problem in a much simpler way, similar to Problem 4.1, in which we used the sum and product of a pair of numbers to find the sum of their reciprocals without ever solving for the numbers themselves.

Problem 6.8

Zoe's mother, Wilma, picks her up at the train station when she comes home from school, and then Wilma drives Zoe home. They always return home at 5:00 p.m. One day, Zoe left school early and got to the train station an hour early. She then started walking home. Wilma left home at the usual time to pick Zoe up and they met along the route between the train station and their house. Wilma picked Zoe up and then drove home, arriving at 4:48 p.m. For how many minutes had Zoe been walking before Wilma picked her up?

Normally, the difficulty with word problems is that there is so much information, and we have to try to parse out the necessary information and organize it into equations. Here, however, the difficulty seems to be that there is hardly any information at all. We don't know how fast the mother drives, how fast Zoe walks, the distance between the house and the station, or when the mother normally leaves the house.

We know only that we are trying to figure out for how long Zoe has been walking. Our first inclination may be to try to use some variation of the formula rate = distance/time. We know that in both scenarios (the scenario in which Zoe gets driven the whole way home and the scenario in which Zoe walks part of the way and gets driven the rest), Zoe travels the same distance—the distance from the train station back to her house. Let's call this distance y.

So let's try to set up two equations of the form rate × time = distance, so that we can equate them to one another. In the first scenario, Zoe is driven the entire way home at a certain rate for a certain amount of time to travel the distance of y. So, our equation will look like this:

$$r_{\text{driving}} \times t = y.$$

In the second scenario, Zoe walks part of the way home and then gets driven the rest of the way, so our equation would look like this:

$$(r_{\text{driving}} \times t_{\text{driving}}) + (r_{\text{walking}} \times t_{\text{walking}}) = y.$$

We have one last clue that we can incorporate here. We know that when Zoe is driven the whole way home, she arrives home at 5:00. However, the day that Zoe walked part of the way, she made it home at 4:48, 12 minutes earlier than usual. She did, however, start her journey an hour earlier than usual, as the problem states, meaning that in total she spent an extra 48 minutes (60 minutes − 12 minutes) on the journey than usual. We cannot be certain of the division of these minutes between driving and walking time, but we at least know that $t_{\text{driving}} + t_{\text{walking}} - 48 = t$, where t is the normal time she takes on her journey. So, we can set our equations equal to each other and then make some substitutions.

$$r_{\text{driving}} \times t_{\text{driving}} + r_{\text{walking}} \times t_{\text{walking}} = r_{\text{driving}} \times t$$

$$r_{\text{driving}} \times t_{\text{driving}} + r_{\text{walking}} \times t_{\text{walking}} = r_{\text{driving}} \times \left(t_{\text{driving}} + t_{\text{walking}} - 48\right)$$

$$r_{\text{driving}} \times t_{\text{driving}} + r_{\text{walking}} \times t_{\text{walking}} = r_{\text{driving}} \times t_{\text{driving}} + r_{\text{driving}} \times t_{\text{walking}} - 48r_{\text{driving}}$$

$$r_{\text{walking}} \times t_{\text{walking}} = r_{\text{driving}} \times t_{\text{walking}} - 48r_{\text{driving}}$$

$$r_{\text{driving}} \times t_{\text{walking}} - r_{\text{walking}} \times t_{\text{walking}} = 48r_{\text{driving}}$$

$$t_{\text{walking}} \times \left(r_{\text{driving}} - r_{\text{walking}}\right) = 48r_{\text{driving}}$$

$$t_{\text{walking}} = \frac{48r_{\text{driving}}}{r_{\text{driving}} - r_{\text{walking}}}.$$

We have now solved for t_{walking} … sort of. If we knew the rate of Wilma's driving and the rate of Zoe's walking, we would be able to solve for the amount of time that Zoe was walking. But how do we get there from here without this information? We've already taken up a lot of time, and now don't know where to go from here. Clearly we need an insight.

Insight 1

Our first insight comes from a complete change of perspective. We stop trying to solve the problem algebraically, and instead approach it geometrically. We will set up a graph with time on the horizontal axis, and location (distance) on the vertical axis as follows:

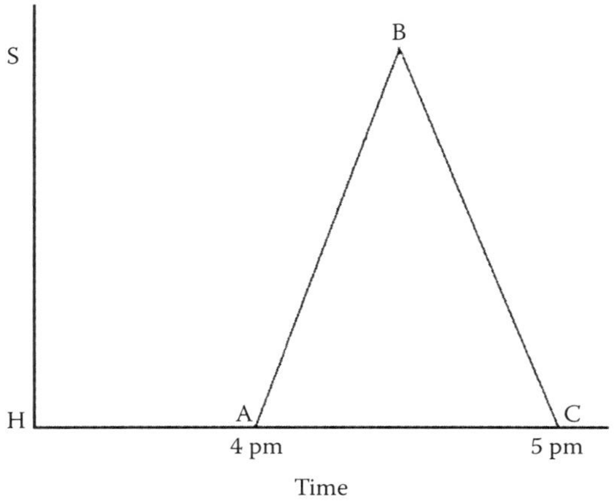

On the vertical axis, we have two locations H for home and S for station. We see that at some arbitrary time, which we call point A, the mother leaves the house and drives at a constant rate toward the station. We don't know that rate, but since it's constant, we know that her trip can be represented by a straight line whose slope corresponds to her driving speed. She then turns around at time point B, drives back home at the same rate, and arrives home at 5 p.m., which is time point C. Thus, we see that her course forms an isosceles triangle.

Now we introduce what happens on the day Zoe gets to the station 1 hour early:

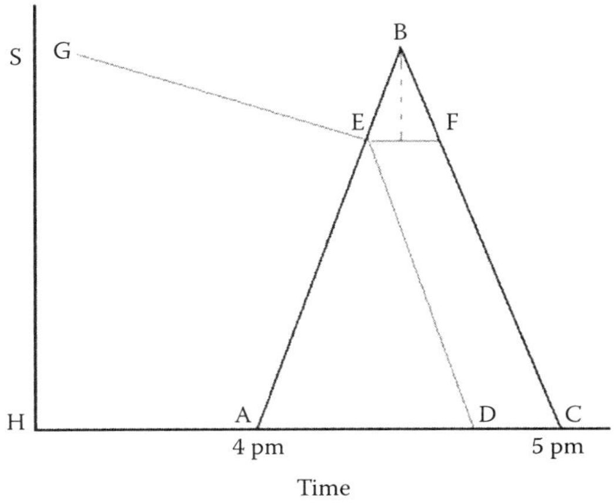

Zoe starts walking at time point G, which we know is 1 hour before time point B, when she normally arrives at the station and gets picked up by her mother. Zoe starts walking toward home, and presumably her walking rate is slower than the driving rate so the slope of her line is shallower than the driving rate. Zoe meets her mother at some point along the mother's way to the station, at point E. She gets into the car, and they start driving home at the usual rate (along segment ED) and get home at point D, which is 12 minutes before point C. The key insights are the following: ED is parallel to BC (because the driving rate is always the same). Also, we can put in horizontal segment EF, so EDCF is a parallelogram, and EF is the same length as DC (i.e., equivalent to 12 minutes). We see that EBF is an isosceles triangle, and we can drop a perpendicular angle bisector from B (because EBF is isosceles) to divide EF in half. Thus, the time difference between E and B is 6 minutes. However, the time Zoe has been walking is the difference between the x-coordinates of G and E. Since the time difference between G and B is 1 hour, and the difference between E and B is 6 minutes, the difference between G and E is 1 hour minus 6 minutes, or 54 minutes. That is how long Zoe has been walking. It is very nice how a graphical approach helped solve the problem so efficiently, despite our lack of so many pieces of information.

Insight 2

There is also another, even easier, way to solve the problem by using an even more clever sort of insight. Our second insight also comes in the form of a change in perspective. It is the natural inclination to look at this problem from Zoe's point of view. However, if we look at this problem from Wilma's point of view, it becomes much easier to solve.

In both scenarios, Wilma leaves to pick up Zoe at the same time as usual. In the second scenario, however, where Wilma meets Zoe on the way, they arrive home 12 minutes earlier than usual, as we already noted. In terms of Wilma's drive, this means that Wilma has saved 6 minutes in each direction of the drive. Therefore, the point where Wilma meets up with Zoe is 6 minutes driving time from the train station, so Wilma picks up Zoe 6 minutes earlier than she usually does. Since Zoe reached the train station and started walking one hour (60 minutes) earlier than usual, but got picked up 6 minutes earlier than usual, we can figure out that Zoe was walking for $60 - 6 = 54$ minutes before she was picked up. Therefore, Zoe was walking for 54 minutes and we have the solution to our problem.

The key to this problem was not algebraic per se, but rather a development in the way we view problems. It is often helpful to try to view problems from multiple directions, because even though a solution may be impossible to find from one angle, it may be readily apparent from another.

Tools Used and Developed

- Algebraic skills and the equation rate = distance/time.
- Recasting an algebra problem into a geometric or graphical representation.
- Viewing a problem from the perspective of a different observer. We can think of this as a change in our "frame of reference" in looking at a problem.

Problem 6.9

A farmer has 400 feet of fence, and needs to fence off a plot of land in the shape of a rectangle to grow tomatoes. What dimensions of the rectangle will maximize his area, and what will the maximum area be?

We start off by making a diagram:

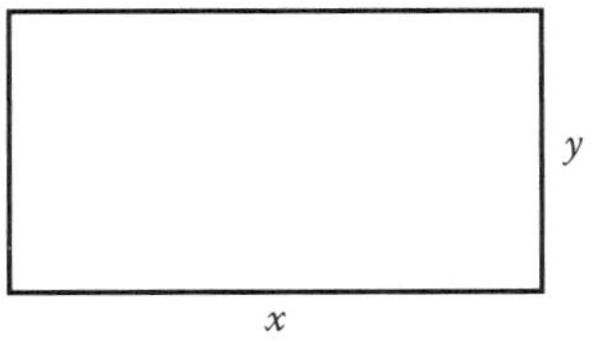

We know that $2x + 2y = 400$, or $x + y = 200$. We wish to maximize the area, xy. There are a couple of ways to go about this (without using calculus). The most direct approach is to use algebra. We know that $y = 200 - x$, so the area $A = xy$ can be written as a function of x as follows:

$$A(x) = x(200 - x) = 200x - x^2 = (-1)(x^2 - 200x).$$

We know that this is an upside-down parabola, so it will have a maximum value. We can find this maximum by completing the square:

$$A(x) = (-1)(x^2 - 200x) = (-1)(x - 100)^2 + 10{,}000.$$

We see that this attains a maximum of 10,000 at $x = 100$ because the term $(x - 100)^2$ is always nonnegative and it gets subtracted from 10,000 (because of the -1 out front). Therefore, the best we can do is to have $(x - 100)^2$ equal 0, which occurs at $x = 100$. Thus, the dimensions that give the maximum area are $x = 100$ and $y = 100$.

This wasn't too bad. It needed a bit of algebra, and knowing how to complete the square. But is there a faster way to get the answer?

Insight 1

We are trying to maximize the quantity xy, given some constraint on $x + y$. It turns out that there is a very nice theorem that can help us do this right away (therefore, this would be a type 2 insight, which would require having seen this theorem before). It is called the arithmetic mean–geometric mean inequality, or the AM-GM inequality for short.

We know how to take the average of two numbers, x and y : $(x + y)/2$. This average is known as the arithmetic mean. There is also another way to take the average, known as the geometric mean: $\sqrt{xy}$. This is done by multiplying both numbers and then taking the square root. It basically asks if a rectangle has dimensions of x and y, what dimensions would give a square of the same area—it is a geometric way of looking at the average of side lengths. What the AM-GM inequality states is that the arithmetic mean is always greater than or equal to the geometric mean: $(x + y)/2 \geq \sqrt{xy}$, with equality occurring only when $x = y$. We can easily prove the AM-GM inequality as follows: we multiply by 2 and square both sides to get $(x + y)^2 \geq 4xy$. This simplifies to $x^2 + 2xy + y^2 \geq 4xy$, or $x^2 - 2xy + y^2 \geq 0$. This is always true because it can be rewritten as $(x - y)^2 \geq 0$, with the left-hand side always being nonnegative. Equality occurs when $x = y$. At this point, the geometric mean would be equal to the arithmetic mean, and we know that this is the maximum value the geometric mean can have.

Therefore, without any calculations, we could have said that if the sum of $x + y$ is fixed (i.e., a fixed perimeter for the fence), the area will be maximized when $x = y$ (because if we maximize $\sqrt{xy}$, we also maximize xy). We know $x + y = 200$, so the maximum area is attained when $x = y = 100$. This leads to a well-known idea in math, that if we have a fixed rectangular perimeter, the area is maximized when the sides of the rectangle are equal, that is, when the shape is a square.

As an aside, the AM-GM inequality generalizes to multiple variables, and it turns out to be true that

$$\frac{x_1 + x_2 + \cdots + x_n}{n} \geq \sqrt[n]{x_1 \cdot x_2 \cdot \cdots \cdot x_n}\,.$$

After warming us up with this problem, let's assume that the farmer has the same 400 feet of fence and wants to fence off a plot to grow tomatoes, but now the back edge of the plot is formed by the wall of the barn. Thus, the fence encloses only three sides of the plot, as shown below. We still want to maximize the planting area.

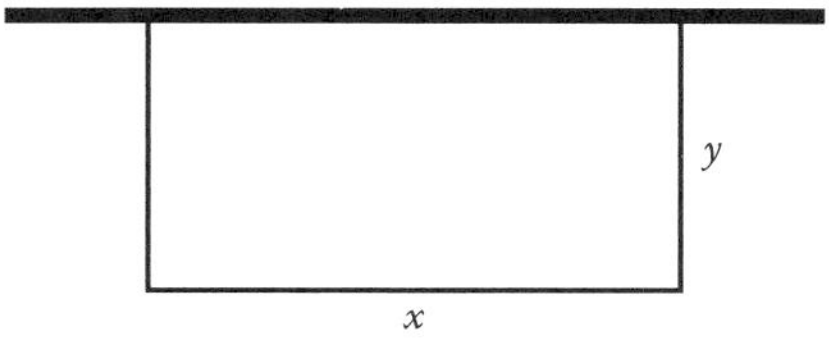

Fresh from the previous problem, the one with four sides of fencing, we immediately jump on the solution. Obviously, the area is maximized when the plot is square shaped, so we will use our fencing to make a square, with $x = y = 133.33$ feet, and the wall of the barn making the back side of the square. Thus, the maximum area will be approximately $(133.33)^2 = 17{,}776.86$ ft^2. This is a nice increase from our prior figure of 10,000, but the answer is incorrect. Let's do the problem the long way, and maybe we can see why that is. We have the following parameters:

$$x + 2y = 400, \quad y = 200 - \frac{x}{2},$$

$$A(x) = xy = x\left(200 - \frac{x}{2}\right) = 200x - \frac{x^2}{2} = \left(-\frac{1}{2}\right)\left(x^2 - 400x\right).$$

Now when we complete the square here, we get

$$A(x) = \left(-\frac{1}{2}\right)\left(x^2 - 400x\right) = \left(-\frac{1}{2}\right)(x - 200)^2 + 20{,}000.$$

We see that this will be a maximum when $x = 200$ (and thus $y = 100$), and that the maximum value will be 20,000, which is indeed bigger than our prior 17,776.89 solution. Surprisingly, if the barn forms the back wall, the area is maximized when the rest of the fence is used to make a rectangle with the width twice the length. Is there an insight that could have saved us from all this?

Insight 2

Once we had solved the first problem, there is indeed a beautiful insight that would have immediately led to the second solution. We showed that if we use all of our fencing to make a square, this maximizes the area. Now imagine that we have double the fencing to 800 feet, and that the barn wall is just a mirror that will reflect whatever structure we make from one side to the other. To maximize the area, we would fashion our 800 feet of fencing into a big square, half of which is reflected through the barn wall, as seen in the following plot.

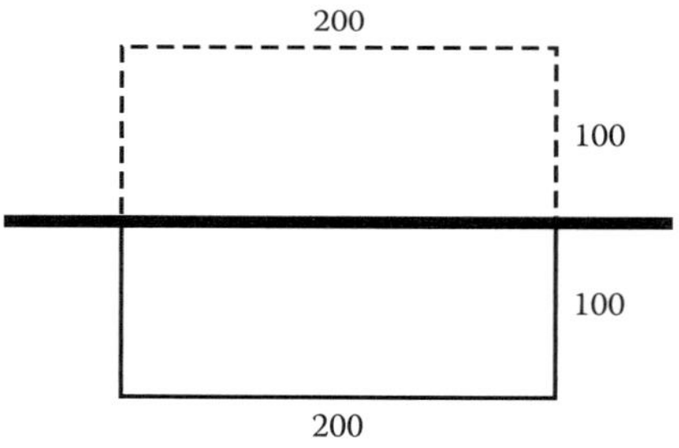

With this insight, and without any further calculation, we could have immediately arrived at the correct dimensions of our plot. What was required here was a change of perspective, to see the barn wall differently: as a mirror and not as a hard barrier. This would allow us to immediately build on our prior solution.

Tools Used and Developed

- Honing our skills to look at problems from a different perspective.
- Completing the square to maximize a function.
- The AM-GM inequality,

$$\frac{x+y}{2} \geq \sqrt{xy}, \quad \frac{x_1+x_2+\cdots+x_n}{n} \geq \sqrt[n]{x_1 \cdot x_2 \cdot \cdots \cdot x_n}.$$

Problem 6.10

Three "quarter circles" and one "three-quarter circle"—all of radius 10—make the attractive jug shape shown below. What is its area?

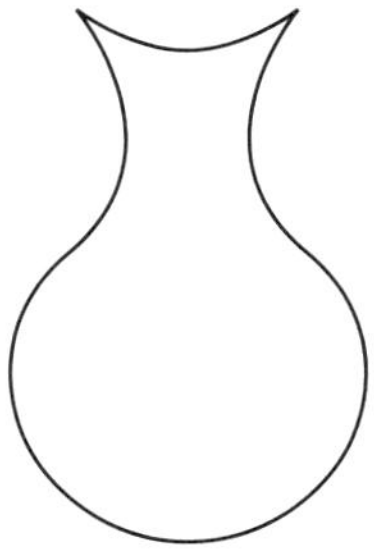

This one is a bit tough. As we start to think about it, we say to ourselves that if we were to complete the circle, we would at least know that the area of the circle is $\pi r^2 = 100\pi$:

But we see that the figure has an area a bit bigger than the circle alone. How do we figure out the area of the top piece? This seems very difficult, as it is not a simple geometric shape. Clearly, we need an insight.

Insight

The key insight here is to look at the problem differently by adding in some elements that transform our perspective. Specifically, let us recall that the shape is made from three quarter circles and one three-quarter circle, each with a radius 10. Let us redraw our diagram, putting in the centers of these circles and joining them up as shown below.

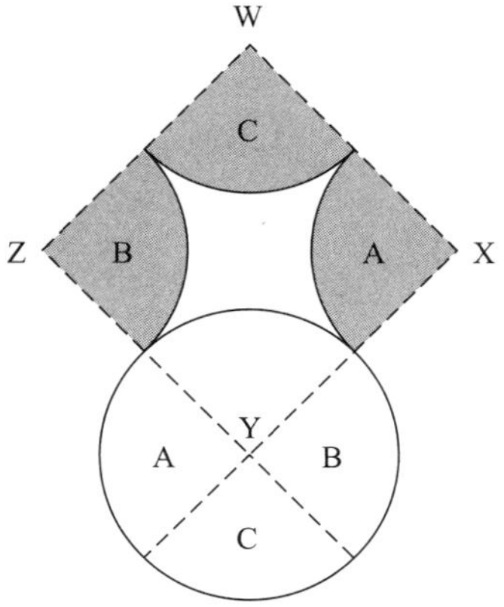

Now we see that we can mentally shift areas A, B, and C in the circle to the shaded sectors A, B, C in the square. These are entirely equivalent, and each represents the area of a quarter circle. Region Y stays where it is, and the troublesome neck of the vase now completes the area of square WXYZ. We see that this square has side length 20 (it is the equivalent of two circle radii), and hence the total area of the figure is 400.

The solution seems amazingly simple once we see it, but requires an ingenious shift in perspective.

Extension

Once we have internalized the above insight, it makes doing other similar problems easier. Let's look at a similar problem, which is to find the area of the figure shown below.

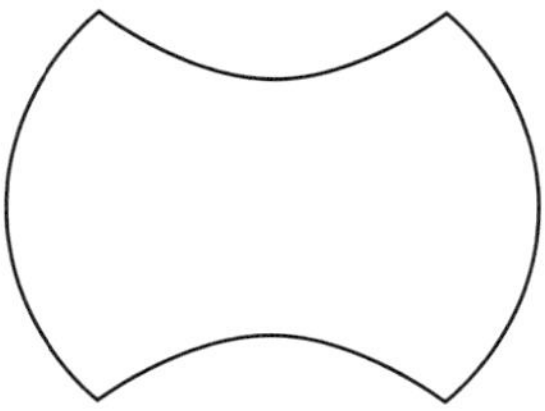

This figure is made up of four quarters of a circle of radius 1 cm arranged as above. In short, the top and bottom quarters have been inverted inward. Once again, we are faced with having to find the area of a strange shape without an easy area formula that we learned in school. Now, though, we are ready. We know that we should add in something to make the problem easier. After a couple of tries, we do the following:

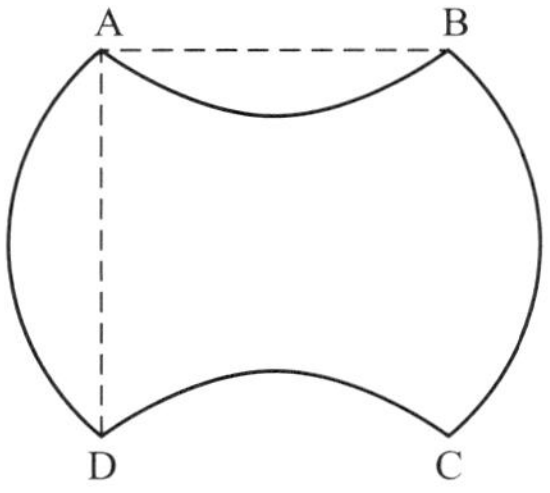

We label points A, B, C and D at the meeting points of the quarter circles, and start to connect them up. We see right away that the bulge outside of line segment AD is precisely equal in area, by symmetry, to the dent beneath segment AB. The same goes for segments BC and CD. Thus, we see that the area of the figure is really the area of square ABCD, which is the in-square of a circle whose diameter is AC. We know that the radius of that circle is 1, so the area is π. From Problem 6.3, we know that the ratio of the area of the in-square to the circumcircle is $2/\pi$, so the area of our figure is $(2/\pi)\pi = 2$. Very easy, this time.

That is one of the points of this book, that insights become easier once we have done similar problems. We may watch someone do a problem and be utterly baffled by their genius, saying to ourselves, "I could have *never* seen that solution." However, maybe that person has done a similar problem or already internalized a similar insight. That would be very similar to how we solved the extension problem after internalizing the insight of the main problem. People watching us solve the extension problem would then be saying to themselves, "Wow, how did he (or she) come up with that?"

Tools Used and Developed

- Perspective shift. In particular, for geometry problems, it is often useful to add in some creative elements to the initial problem.

7

Miscellaneous Challenging Problems

Problem 7.1

The numbers 1, 2, 3, ..., n^2 are organized into an $n \times n$ matrix M as follows:

$$M = \begin{pmatrix} 1 & 2 & \dots & n \\ n+1 & n+2 & \dots & 2n \\ \dots & \dots & \dots & \dots \\ (n-1)n+1 & (n-1)n+2 & \dots & n^2 \end{pmatrix}.$$

From this matrix, a sum S is generated as follows: pick some element x_1 at random, then delete the remaining entries in its corresponding row and column. From the remaining entries, pick a second element x_2, and then delete the remaining elements in its row and column, and continue this procedure until the matrix M is exhausted. S is the sum $x_1 + x_2 + \dots$. Will the sum S always be constant no matter how the elements x_1, x_2 ... are chosen? If so, what is the value of this sum?

This seems a bit difficult to answer, so we'll start with a small case to get ourselves oriented. Let $n = 3$. Thus, the numbers 1, 2, 3, ..., 9 are organized in a 3×3 matrix M:

$$M = \begin{pmatrix} 1 & 2 & 3 \\ 4 & 5 & 6 \\ 7 & 8 & 9 \end{pmatrix}.$$

Let us pick an element from this matrix, say, 6. So, $S = 6 + \ldots$. We now delete the row and column that contain 6 and we have

$$M = \begin{pmatrix} 1 & 2 & - \\ - & - & - \\ 7 & 8 & - \end{pmatrix}.$$

Now we pick 8 and repeat the same procedure by deleting the remnants of the third row and the second column, so we now have $S = 6 + 8 + \ldots$:

$$M = \begin{pmatrix} 1 & * & - \\ - & - & - \\ * & * & - \end{pmatrix}.$$

The only element left now is 1, so $S = 6 + 8 + 1 = 15$.

Let us try again, with a different starting value—let's pick 9 this time.

$$M = \begin{pmatrix} 1 & 2 & 3 \\ 4 & 5 & 6 \\ 7 & 8 & 9 \end{pmatrix}.$$

So, $S = 9 + \ldots$, and the matrix M becomes

$$M = \begin{pmatrix} 1 & 2 & - \\ 4 & 5 & - \\ - & - & - \end{pmatrix}.$$

Next, we pick 4, and deleting the remnants of the second row and the first column, we see that the 2 would be left, so $S = 9 + 4 + 2 = 15$.

It seems that perhaps S might be constant and its value may be 15. However, this is far from mathematical proof. To prove the case for our small 3×3 matrix, we would need to try all possible sums. But how many of those are there, anyway? Just answering this preliminary question seems quite complex in itself and still doesn't solve our problem. However, assuming that $S = 15$, how does this relate to n? If we try to generalize, could $S = 5n$? Or perhaps $S = n^2 + 2n$? Or perhaps $S = 2n^2 - n$?

Obviously, this sort of guessing is neither productive nor conclusive. Even if we were to somehow guess correctly, we would have no way to verify this. To continue, we will need some sort of insight.

Insight

To begin with, we note that if we pick an element a_i from matrix M, we delete its row. Thus, all of the elements will come from different rows. In other words, S will have n terms, each from a different row. The same applies to columns, so S will have n terms, no two of which are from the same row or the same column.

Let us call a_i the term in S from the ith row of M. Thus, a_1 is the element in S that came from the first row (not the element that was chosen first). Thus, we can see that

$$S = a_1 + a_2 + a_3 + \ldots + a_n,$$

although the elements most likely were not picked in that order. For example, the contributions to S may have come in the following way

$$M = \begin{pmatrix} & & & a_1 \\ a_2 & & & \\ \cdots & \cdots & \cdots & \cdots \\ & a_n & & \end{pmatrix}.$$

Let's make another $n \times n$ matrix N, each of whose rows are the integers 1, 2, …, n. Thus, N is

$$N = \begin{pmatrix} 1 & 2 & \cdots & n \\ 1 & 2 & \cdots & n \\ \cdots & \cdots & \cdots & \cdots \\ 1 & 2 & \cdots & n \end{pmatrix}.$$

Namely, the first column of N consists of all 1s, the second column consists of all 2s, and the ith column consists of all is.

Now let's pick elements b_i from N that correspond precisely to the locations of the elements a_i from M that went into the sum S. Thus, in our example, we have

$$M = \begin{pmatrix} & & & a_1 \\ a_2 & & & \\ \cdots & \cdots & \cdots & \cdots \\ & a_n & & \end{pmatrix}, \quad N = \begin{pmatrix} & & & b_1 \\ b_2 & & & \\ \cdots & \cdots & \cdots & \cdots \\ & b_n & & \end{pmatrix}.$$

Matrices M and N have the same first row. In the second row, each term in M is n more than each corresponding term in N. In the third row, each term in M is $2n$ more than each corresponding term in N; similarly, in the ith row, each term in M is $(i - 1)n$ more than each term in N. So we can then write

$$a_1 = b_1, \quad a_2 = b_2 + n, \quad a_3 = b_3 + 2n, \quad \ldots, \quad a_n = b_n + (n - 1)n.$$

With this information in hand, we can write S as

$$\begin{aligned} S &= a_1 + a_2 + a_3 + \ \ldots \ + a_n \\ &= b_1 + (b_2 + n) + (b_3 + 2n) + \ \ldots. \ + b_n + (n-1)n. \end{aligned}$$

By regrouping and factoring out an n, we can rewrite S as

$$S = (b_1 + b_2 + b_3 + \ldots. + b_n) + n[1 + 2 + 3 + \ldots. + n - 1].$$

However, we remember that the b_i terms come one from each column of N, so they have to be some permutation of 1, 2, 3, … , n. Thus,

$$b_1 + b_2 + b_3 + \ldots. + b_n = 1 + 2 + 3 + \ldots. + n.$$

Now we are on familiar territory. Using the summation formula for the first n natural numbers that we developed in Problem 1.1,

$$1 + 2 + 3 + \ \ldots . + n = \frac{n(n+1)}{2},$$

we have

$$b_1 + b_2 + b_3 + \ \ldots . + b_n = \frac{n(n+1)}{2}.$$

Also, using the same formula and substituting $n - 1$ for n, we have

$$1 + 2 + 3 + \ \ldots . + n - 1 = \frac{(n-1)n}{2}.$$

Thus, we can write S as

$$\begin{aligned} S &= (b_1 + b_2 + b_3 + \ \ldots . + b_n) + n[1 + 2 + 3 + \ \ldots . + (n-1)] \\ &= \frac{n(n+1)}{2} + n\frac{(n-1)n}{2} \\ &= \frac{n}{2}(n + 1 + n^2 - n) \\ &= \frac{n(n^2+1)}{2}. \end{aligned}$$

Thus, we see that S is a constant, dependent only on n, the size of the matrix, and completely independent of the manner in which we pick the elements that go into S.

Returning to our simple example where $n = 3$, we see that $S = 3(3^2 + 1)/2 = 15$, just as we got in our few trial-and-error attempts. Now we understand why and how we got it!

As advertised, we are now solving Olympiad-level problems by using just the simple tools we have developed thus far.

Tools Used and Developed

- The summation formula for the first n natural numbers, $1 + 2 + 3 + \ldots . + n = n(n + 1)/2$, which we developed in Problem 1.1.

Problem 7.2

Evaluate the following fraction:

$$\frac{(2^3-1)(3^3-1)(4^3-1)\ldots(10^3-1)}{(2^3+1)(3^3+1)(4^3+1)\ldots(10^3+1)}.$$

The problem is somewhat tedious, but there is no mystery about how to do it. We evaluate each of the parentheses, take the products, get values for the numerator and the denominator, and then simplify the resulting fraction:

$$\frac{(7)(26)(63)\ldots(999)}{(9)(28)(65)\ldots(1{,}001)}.$$

After a remarkable amount of tedious arithmetic, and hoping we made no mistakes, we get 37/55.

Now let's do the same problem using 100 instead of 10:

$$\frac{(2^3-1)(3^3-1)(4^3-1)\ldots(100^3-1)}{(2^3+1)(3^3+1)(4^3+1)\ldots(100^3+1)}.$$

At this point, our earlier approach is clearly too much. We need an insight.

Insight

There are actually a couple of critical insights needed to solve this problem, as is typical of difficult (Olympiad-level) problems, but some of these are tools we have already developed. First, we notice that the terms in the numerator are all of the form $(x^3 - 1)$, and the factors in the denominator are all of the form $(x^3 + 1)$. We may not see how this is immediately helpful, but we remember that differences of cubes and sums of cubes can be conveniently factored:

$$(x^3 - 1) = (x - 1)(x^2 + x + 1),$$

$$(x^3 + 1) = (x + 1)(x^2 - x + 1).$$

Thus, each term in the numerator and the denominator may be rewritten in the above factored form. For example, $(100^3 + 1) = (100 + 1)(100^2 - 100 + 1)$, and $(99^3 - 1) = (99 - 1)(99^2 + 99 + 1)$.

Initially, this does not seem to help—now we have twice as many terms to evaluate. The way out of our dilemma is to notice that although there are many terms, if we could get some cancellations (i.e., through multiplicative telescoping), we would be in great shape. We start by asking whether some of these terms would equal each other. For example, can $(a^3 - 1) = (b^2 - b + 1)$ for some values of a and b between 2 and 99 to get us some cancellations? If this happens, it certainly would be rare. The key lies in the following insight.

Let us focus on the second part of the factorizations, and define two functions, $f(x)$ and $g(x)$ as $f(x) = x^2 + x + 1$, and $g(x) = x^2 - x + 1$. Here is the critical insight. These functions can be expressed in terms of each other as

$$g(x) = f(x - 1).$$

That may not seem too profound, but it is. Let's verify it. The notation $f(x - 1)$ simply means to plug in $(x - 1)$ for x in $f(x)$. Thus,

$$\begin{aligned} f(x-1) &= (x-1)^2 + (x-1) + 1 \\ &= (x^2 - 2x + 1) + (x-1) + 1 \\ &= x^2 - x + 1 \\ &= g(x). \end{aligned}$$

This means, for example, that $99^2 + 99 + 1 = 100^2 - 100 + 1$ (both equal 9,901). Now we can see daylight!

$$\frac{(2^3-1)(3^3-1)(4^3-1)\ldots(100^3-1)}{(2^3+1)(3^3+1)(4^3+1)\ldots(100^3+1)}$$

can be rewritten as

$$\frac{\begin{array}{l}(2-1)(2^2+2+1)(3-1)(3^2+3+1)(4-1)(4^2+4+1)\ldots \\ \qquad\ldots(99-1)(99^2+99+1)(100-1)(100^2+100+1)\end{array}}{\begin{array}{l}(2+1)(2^2-2+1)(3+1)(3^2-3+1)(4+1)(4^2-4+1)\ldots \\ \qquad\ldots(99+1)(99^2-99+1)(100+1)(100^2-100+1)\end{array}}.$$

We can see now where this is going. We see that $(2^2 + 2 + 1)$ in the numerator is equal to $(3^2 - 3 + 1)$ in the denominator because, as we established, $f(x - 1) = g(x)$. Thus, they would cancel each other out. By the same token, $(3^2 + 3 + 1)$ in the numerator is equal to $(4^2 - 4 + 1)$ in the denominator, and they would likewise cancel. This continues through until we get to $99^2 + 99 + 1 = 100^2 - 100 + 1$.

Thus, we get a multiplicative telescoping, where each quadratic term in the numerator cancels the quadratic term to its right in the denominator. Likewise, the linear term $(4 - 1)$ in the numerator cancels the $(2 + 1)$ term in the denominator, the $(5 - 1)$ term in the numerator cancels the $(3 + 1)$ term in the denominator, through the $(100 - 1)$ term in the numerator, which cancels the $(98 + 1)$ term in the denominator.

Thus, the product reduces to

$$\frac{(2-1)(3-1)(100^2+100+1)}{(2^2-2+1)(99+1)(100+1)} = \frac{2(10{,}101)}{3(100)(101)} = \frac{3{,}367}{5{,}050}.$$

Tools Used and Developed

- Multiplicative telescoping.

Problem 7.3

Calculate the number of internal intersections made by the diagonals of irregular convex polygons.

We don't want to get caught up in jargon. The part about "convex" just means that when we connect any two vertices, the line is contained within the polygon, so we are not talking about things like a star. When the polygon is convex, all intersections occur inside the polygon. The part about "irregular," means that only two diagonals intersect at any point. This actually makes the problem much easier. If the polygon is regular, then more than two diagonals can meet at a point, and that makes the problem much harder. So hard that it becomes a problem for professional mathematicians. In fact, that harder problem was solved by two excellent MIT mathematicians, Bjorn Poonen and Michael Rubinstein (Poonin and Rubenstein, 1998).

Now, getting back to our easier problem, we can just take it at face value. Imagine that we have an irregular polygon, made by putting some unevenly spaced points on a circle and connecting them. If we connect all the vertices to each other, then we have diagonals that intersect. How many intersections are there? To make things more concrete, look at the irregular hexagon ABCDEF below. How many internal intersections are there?

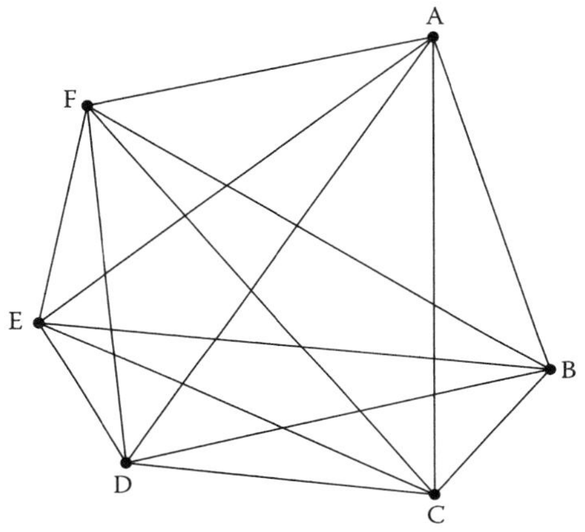

First Solution: Brute Force

The first way to do the problem is by brute force. We just count the intersections inside the hexagon. Even for such a small number of vertices ($n = 6$), this gets a little confusing. Try it! I miscounted twice, and then finally got 15. However, we can see that this method will quickly become impractical. For example, if we are dealing with a nonagon ($n = 9$), it becomes nearly impossible (see the figure below), let alone for huge values of n.

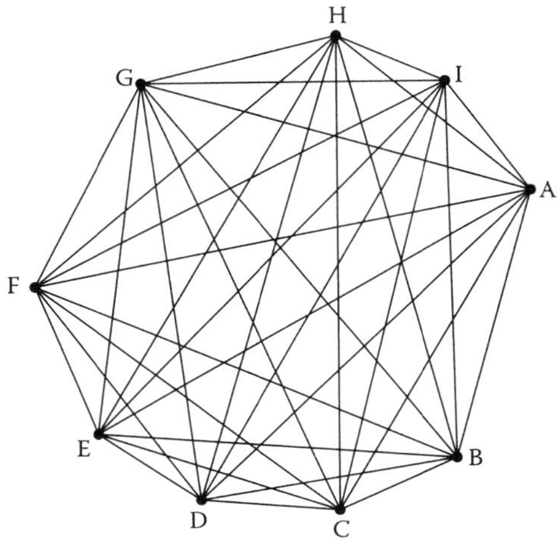

Second Solution: Solving by Iteration

We can be a bit more creative and seek a general solution by "building it up" from smaller cases. This is known as "iteration." By using iteration, we can get a summation relationship for the number of intersections by building an n-gon from an $(n - 1)$-gon, by adding a new vertex point and seeing what happens. First, we note that as an $(n - 1)$-gon is converted to an n-gon, the new diagonals add new intersection points without affecting any of the intersection points already present in the $(n - 1)$-gon. This is a key insight. The question becomes one of finding an expression for these new intersections in terms of n. This initially seems very complex, but will quickly become clear if we look at a concrete example, such as that shown below.

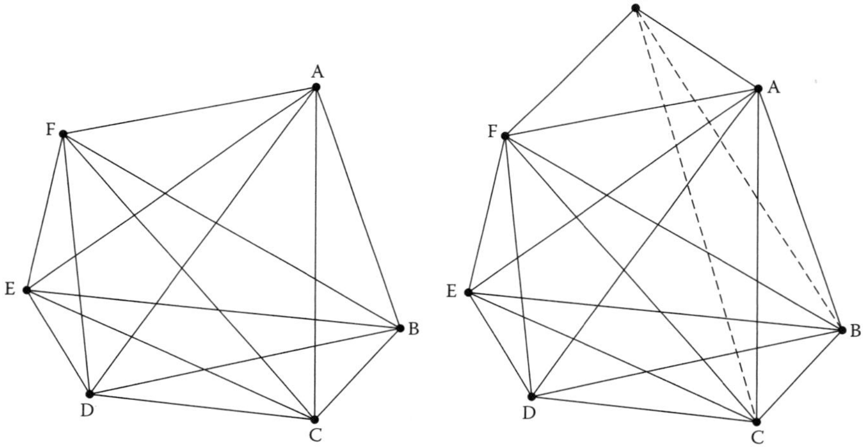

This figure shows our original hexagon ABCDEF and its conversion to a heptagon by adding another vertex point, G. Vertex G is connected by sides to vertex F and vertex A. Vertex G will now send four new diagonals to points B, C, D, and E. The first diagonal, GB, divides the heptagon into two parts. On one side is point A, and on the other side are points C, D, E, and F. These two sets, {A} and {C, D, E, F} form four diagonals (AC, AD, AE, and AF), all of which intersect GB. Therefore, GB produces four new intersections. Proceeding clockwise, and considering the next diagonal, GC, we see that it divides the heptagon into two parts, one containing vertices {A, B} and the other containing vertices {D, E, F}. The two sets of vertices, located on "opposite" sides of the heptagon, can form $2 \times 3 = 6$ new diagonals (AD, AE, AF, BD, BE, and BF), all of which intersect GC. Therefore, diagonal GC produces six new intersections. Diagonal GD will similarly produce six new intersections, and finally diagonal GE will produce four new intersections, and the total number of new intersections produced by adding a seventh vertex to the hexagon is the sum of (1 × 4), (2 × 3), (3 × 2), and (4 × 1). This amounts to 20 new intersections, to be added to the ones already in the hexagon. Since the hexagon had 15 intersections, the heptagon must have 35.

Furthermore, this example now makes the general pattern apparent. Let us consider that our irregular convex polygons are formed by putting vertex points on the circumference of a circle, and moving sequentially clockwise from vertex X_1 to vertex X_2 … to X_n. When vertex X_n is added to the $(n - 1)$-gon $X_1 - X_{n-1}$, it will connect to points X_1 and X_{n-1}, and send out $n - 3$ new diagonals to points $X_2 \ldots X_{n-2}$. Any one of these diagonals, say, X_nX_k, divides the remaining $n - 2$ vertices of the n-gon into two sets $\{X_1 \ldots X_{k-1}\}$ and $\{X_{k+1} \ldots X_{n-1}\}$ on opposite sides of that diagonal. Each pair of vertices, one chosen from each set, makes an internal diagonal that intersects X_nX_k, for a total of $(k - 1)(n - k - 1)$ new intersections due to that diagonal. Diagonals formed from pairs of vertices in the same set (such as X_2X_{k-2}) will not intersect diagonal X_nX_k. If we look sequentially at all of the $n - 3$ diagonals, moving clockwise from X_n, we see that they form sets whose cardinalities are all of the nonzero two-part partitions of $n - 2$. The total number of new intersections, therefore, is the sum of the products of these two-part partitions. This can be more easily visualized by looking at the grid below.

	1	2	3	4	5	6
1	1	2	3	4	5	6
2	2	4	6	8	10	12
3	3	6	9	12	15	18
4	4	8	12	16	20	24
5	5	10	15	20	25	30
6	6	12	18	24	30	36

The cells in the grid are the product of the row and column in which they occur. Each diagonal trace shows the new intersections produced by adding another vertex point to a triangle, which has no intersections. Thus, adding the first vertex point, to make a quadrilateral, produces 1 new intersection.

Adding a second point to make a pentagon produces 2 + 2 new intersections on top of that. We notice that these new intersections are the product of 1 × 2 and 2 × 1, or the two ways we can divide 3 into two parts. When we add on a sixth point to make the hexagon, we add the products of the two-part cardinalities of 4 (1 × 3, 2 × 2, and 3 × 1), or the third diagonal trace in our grid, and so on. When we made the heptagon from the hexagon in our concrete example above, we produced (1 × 4), (2 × 3), (3 × 2), and (4 × 1) (i.e., 4 + 6 + 6 + 4) new intersections, which we now see as the fourth diagonal trace in our grid. We can easily predict that adding a 9th vertex point to form a 9-gon (nonagon) from an octagon will produce new intersections whose number is the sum of the shaded cells along the diagonal of our grid. To get the total number of intersections within the nonagon, we would sum all the diagonal traces above and including the shaded diagonal. These total 126. Thus, the internal intersections within a convex polygon formed by adding k additional vertices to our starting triangle will be given by the left upper triangular sum, including the "anti-diagonal," of the elements of our $k \times k$ grid, where each element is the product of its row and column, and where the kth diagonal trace represents the products of the nonzero two-part partitions of $k + 1$. Thus, for an n-gon, we would let $k = n - 3$ (e.g., our 6 × 6 grid above gives the total for a 9-gon).

Third Solution: The "Aha" Solution

This one-step solution depends on the insight that for an irregular convex n-gon, any four vertices are the endpoints of a unique pair of intersecting diagonals, and hence determine a unique intersection point. If we pick four different vertices, they determine two different diagonals, and hence a different intersection point. Therefore, the total number of intersection points is the number of ways four vertices can be chosen from the n vertices of the n-gon, or $\binom{n}{4}$, which means $n!/[(n-4)!4!]$. Thus, for $n = 6$, this means that the number of internal intersection points is $6!/(2!4!) = 15$. For $n = 7$, it is $7!/(3!4!) = 35$, and for $n = 9$, it is $9!/(4!5!) = 126$, just as we got with our brute force and iteration methods. See how much work one key insight can save?

Now that we have this insight, we can capitalize on it to solve a related problem. Our polygons get divided into multiple regions by the intersecting diagonals. How many internal triangles will there be in an n-gon? "Internal triangles" here means that all the vertices of the triangle lie inside the n-gon (i.e., no vertex of the triangle is a vertex of the n-gon). The solution to this problem appeared in *Mathematics Magazine* (Moser, 1953), so we are now solving high-level problems.

Each triangle is uniquely made by three diagonals, each of which is determined by a pair of vertices of the n-gon. Conversely, every triangle has sides that lead to a unique set of six vertices of the n-gon. Therefore, the number of internal triangles in $C(n,6)$ or $n!/[6!(n-6)!]$. Thus, for a hexagon, we expect only one internal triangle, whereas for a heptagon we expect seven.

Tools Used and Developed

- Iteration to set up a problem.
- The internal intersections of irregular convex n-gons is given both by the sum of the products of the nonzero two-part partitions filling the $k \times k$ grid shown in this solution, where $k = n - 3$, and by the simple formula $\binom{n}{4}$.

Problem 7.4

Recall the well-known Christmas carol, "The Twelve Days of Christmas," the first part of which is given below. Over the twelve days of Christmas, how many gifts did my true love give to me?

On the first day of Christmas,
my true love gave to me
A partridge in a pear tree.
On the second day of Christmas,
my true love gave to me
Two turtle doves,
And a partridge in a pear tree.
On the third day of Christmas,
my true love gave to me
Three French hens,
Two turtle doves,
And a partridge in a pear tree.
On the fourth day of Christmas,
my true love gave to me
Four calling birds,

Three French hens,
Two turtle doves,
And a partridge in a pear tree.
....

Brute Force Solution

We can just make a table for the number of gifts received each day, and add them up.

Day	*Gifts*	*Day's Total*
1	1	1
2	1+2	3
3	1+2+3	6
4	1+2+3+4	10
5	1+2+3+4+5	15
6	1+2+3+4+5+6	21
7	1+2+3+4+5+6+7	28
8	1+2+3+4+5+6+7+8	36
9	1+2+3+4+5+6+7+8+9	45
10	1+2+3+4+5+6+7+8+9+10	55
11	1+2+3+4+5+6+7+8+9+10+11	66
12	1+2+3+4+5+6+7+8+9+10+11+12	78

If we sum these, we get a total of 364 gifts. Of course, we could have used our familiar formula to help sum each row, $1 + 2 + 3 + \ldots + n = n(n + 1)/2$

We can see that as this problem generalizes to n days of Christmas, this brute force approach will become quite involved. An insight will help to figure another way to do it.

Insight

Let us start as we did before, summing up each day's gifts by using the formula for summing the integers $1 \ldots n$. Let us start from day n, and work backward. We would have the following:

$$\frac{n(n+1)}{2}+\frac{(n-1)n}{2}+\frac{(n-2)(n-1)}{2}+\frac{(n-3)(n-2)}{2}+\ldots+\frac{2(3)}{2}+\frac{1(2)}{2}.$$

The first term, $n(n + 1)/2$, represents the number of gifts we got on day n; the next term, $(n - 1)n/2$, represents the number of gifts we got on day $n - 1$, and so on. The last two terms represent the number of gifts we got on day 2 and day 1, respectively.

To simplify this sum, let us use the insight that we can group and sum the terms in groups of 2. This, by the way, is often a very useful tool. In this way, we can factor and get a simpler expression. Thus, we can look at the sum in this way:

$$\left[\frac{n(n+1)}{2}+\frac{(n-1)n}{2}\right]+\left[\frac{(n-2)(n-1)}{2}+\frac{(n-3)(n-2)}{2}\right]+\ldots+\left[\frac{3(2)}{2}+\frac{2(1)}{2}\right].$$

Let us look at the first set of brackets, the sum for day n and day $n-1$. We can simplify the sum by factoring out an n, noticing the beautiful cancellation that occurs when we sum $(n+1)$ and $(n-1)$. Thus, we can rewrite the sum as

$$\left[\frac{n(n+1)}{2}+\frac{(n-1)n}{2}\right]=\frac{n(\,(n+1)+(n-1)\,)}{2}=\frac{n(2n)}{2}=n^2.$$

Let's sum the next bracket, for days $n-2$ and $n-3$. We see that we can factor out an $(n-2)$, and again we get a nice simplification:

$$\left[\frac{(n-2)(n-1)}{2}+\frac{(n-3)(n-2)}{2}\right]$$

$$=\frac{(n-2)(\,(n-1)+(n-3)\,)}{2}=\frac{(n-2)(2(n-2))}{2}=(n-2)^2.$$

Summing day 2 and day 1 in a similar fashion, we have,

$$\left[\frac{3(2)}{2}+\frac{2(1)}{2}\right]=2^2.$$

Thus, we see the pattern. When we sum in groups of two, the sum collapses to summing every other square. For an even number of days, we have

$$n^2+(n-2)^2+(n-4)^2+\ldots+2^2.$$

For example, for our problem, the total now becomes $12^2 + 10^2 + 8^2 + 6^2 + 4^2 + 2^2$.

Let's continue with the general case, $n^2 + (n-2)^2 + (n-4)^2 + \ldots + 2^2$. Since n is even in our case, we can say that $n = 2k$. Thus, $n - 2$ would be $2(k-1)$, $n - 4$ would be $2(k-2)$, and so on. Our sum then becomes

$$n^2 + (n-2)^2 + (n-4)^2 + \ldots + 2^2$$
$$= (2k)^2 + \left(2(k-1)\right)^2 + \left(2(k-2)\right)^2 + \ldots + (2(1))^2.$$

We can factor out a 2^2 from each term and get our sum as

$$4\left[(k)^2 + (k-1)^2 + (k-2)^2 + \ldots + (1)^2\right].$$

Using summation notation, we can rewrite this as:

$$4\sum_{i=1}^{k} i^2,$$

where $k = n/2$. Thus, for our 12-day problem, the answer becomes $4[1^2 + 2^2 + 3^2 + 4^2 + 5^2 + 6^2]$.

Let us return to the general case. In Problem 1.14, we had gone to the trouble of evaluating the sum of the first k squares,

$$\sum_{i=1}^{k} i^2 = \frac{k(k+1)(2k+1)}{6}.$$

Now we can use this tool. The answer to our problem, where n is an even number of days, is

$$4\,\frac{k(k+1)(2k+1)}{6} = \frac{2k(k+1)(2k+1)}{3},$$

where $k = n/2$. Thus, for our 12 days of Christmas, we have $n = 12$, and $k = 6$. Thus, the number of gifts is:

$$\frac{2(6)(6+1)(2\times 6+1)}{3} = 13\times 28 = 364,$$

just as we obtained before using our brute force counting. Moreover, we now have a general formula. With more manipulating of the sum, we can also derive a formula for n odd. However, instead of expending the time on that, we will go after an even more keen insight.

A More Keen Insight

In our prior two approaches, we focused on the number of gifts obtained each day, and found a clever way to total them. Often, a change of perspective yields powerful new insights. Let us instead focus on the actual gifts, and sum across the type of gift. For example, each day, we received a single partridge in a pear tree, so we have 1×12 of those across the entire period. In terms of turtle doves, we received two of those on the second day, and every day thereafter, so we have a total of 2×11 turtle doves. On the third day, we received three French hens, and continued to receive three of them for the rest of the 12-day period, so we got a total of 3×10 French hens. Continuing the pattern, we see that we got 4×9 calling birds. We can now see the pattern for the total number of gifts, summed by the type of gift:

$$S = (1 \times 12) + (2 \times 11) + (3 \times 10) + (4 \times 9) + \ldots.. + (11 \times 2) + (12 \times 1).$$

A big light bulb should have gone on in our heads. We see a familiar pattern. The total number of gifts is the sum of the products of the 2-part partitions of 13. We encountered just this sort of pattern in Problem 7.3. It would be one of the diagonals in the matrix we used to tally the number of internal intersection points in an irregular convex polygon. Very specifically, for the 2-part partitions of 13, the sum above represents the new intersections added by the 15th vertex point, as we go from a 14-gon to a 15-gon, as we derived before. That is absolutely amazing! There is a hidden relationship between two problems that would seem to have absolutely nothing at all to do with each other.

How do we find the total of the new intersections added by the 15th vertex point of the 15-gon? It would be the total number of internal intersections in a 15-gon minus the number of intersections in a 14-gon. That subtraction would yield the number of new intersections given by adding the 15th vertex point to convert a 14-gon to a 15-gon.

As we said before, the number of internal intersections in an irregular convex n-gon is just $\binom{n}{4}$. Hence, for our problem, the answer would be

$$S=\binom{15}{4}-\binom{14}{4}.$$

Now we have a one-step solution to our problem. However, let us tidy up this solution a bit. From the theory of combinatorics and Problem 2.6, we know a very interesting fact about an equation of the sort $\binom{n}{k}-\binom{n-1}{k}$ given by something called Pascal's identity, which states that $\binom{n-1}{k}+\binom{n-1}{k-1}=\binom{n}{k}$. This identity can be proven by direct substitution. It is a very useful identity in combinatorics.

Thus, applying Pascal's identity to $S = \binom{15}{4}-\binom{14}{4}$, we get that $S = \binom{14}{3}$. When we calculate this out, we get

$$S=\binom{14}{3}=\frac{14!}{3!(11!)}=364.$$

We have truly derived a one-step solution to our problem, using a most-unexpected connection with the problem of the internal intersections formed by the diagonals of irregular convex polygons. We need to clarify this solution just a bit because the indexing could get very confusing. We remember from Problem 7.3 that the number of new intersections added by the nth vertex point is actually the sum of the products of the two-part partitions of $(n - 2)$. Thus, the number of new intersections added by the 15th vertex point is the sum of the products of the two-part partitions of 13 given by $\binom{15}{4}-\binom{14}{4}=\binom{14}{3}$. This corresponds to the total number of gifts received over 12 days of Christmas. Thus, generalizing this approach over n days of Christmas, we expect to receive the same number of gifts as the number of new intersections obtained by the $(n + 3)$rd vertex point in an $(n + 3)$-gon, which is given by $\binom{n+3}{4}-\binom{n+2}{4}=\binom{n+2}{3}$. Thus, we have our general formula for the number of gifts received over n days of Christmas as $S = \binom{n+2}{3}$, a neat one-step solution.

One Final Push

Before leaving this problem, let us make one last push to put together the insights that we have gained. We derived that for an even number of days, the total number of gifts is given by $n^2 + (n-2)^2 + (n-4)^2 + \ldots + 2^2$; for example, for the 12 days of Christmas, the total number of gifts is given by $12^2 + 10^2 + 8^2 + 6^2 + 4^2 + 2^2$. We now know that this is equivalent to $\binom{n+2}{3}$, or in this case, $\binom{14}{3}$.

If we return to our prior solution but deal with the problem of an odd number of days, n, we see that if we use the same pairing tool with an odd number of days, we would have paired things into groups of two, and had a 1 left over at the end. In other words, we would have had

$$\left[\frac{n(n+1)}{2}+\frac{(n-1)n}{2}\right]+\left[\frac{(n-2)(n-1)}{2}+\frac{(n-3)(n-2)}{2}\right]+\ldots+$$
$$\left[\frac{4(3)}{2}+\frac{3(2)}{2}\right]+\frac{2(1)}{2}.$$

This would simplify to $n^2 + (n-2)^2 + (n-4)^2 + \ldots + 3^2 + 1^2$, where we have replaced the final 1 by its equivalent, 1^2. Thus, for 13 days of Christmas, the total number of gifts would be $13^2 + 11^2 + 9^2 + 7^2 + 5^2 + 3^2 + 1^2$, still the sum of every other square. We know that this total is also given by our formula, $S = \binom{n+2}{3}$, in this case, $\binom{15}{3}$.

When we add these two results together, we have:

$$\left(13^2+11^2+9^2+7^2+5^2+3^2+1^2\right)$$
$$+\left(12^2+10^2+8^2+6^2+4^2+2^2\right)=\binom{15}{3}+\binom{14}{3}.$$

But the left-hand side is just the sum of *all* the squares from 1 to 13. Thus, we have derived a wonderful new formula for the sum of the first n squares:

$$\sum_{i=1}^{n} i^2 = \binom{n+2}{3}+\binom{n+1}{3}.$$

How does this jibe with our prior formula,

$$\sum_{i=1}^{n} i^2 = \frac{n(n+1)(2n+1)}{6}?$$

We know they have to be the same, and we have proven it by using yet another new tool: counting the same thing in two different ways. A gruesome amount of algebra confirms that both formulas are equal to the same thing:

$$\frac{n^3}{3} + \frac{n^2}{2} + \frac{n}{6}.$$

Tools Used and Developed

- The sum of the first n integers tool, $1 + 2 + 3 + \ldots + n = n/(n + 1)/2$, developed in Problem 1.1.
- The pairing tool and the sum of the first n squares tool,

$$\sum_{i=1}^{n} i^2 = n(n+1)(2n+1)/6,$$

  derived in Problem 1.14, to obtain a closed-form solution.
- Tools derived from the internal intersections of a convex n-gon, that the number of new intersections added by the nth vertex is simultaneously the sum of the products of the two-part partitions of $(n - 2)$, and also given by the formula for $\binom{n}{4} - \binom{n-1}{4}$, all of which was derived in Problem 7.3.
- Pascal's identity, $\binom{n-1}{k} + \binom{n-1}{k-1} = \binom{n}{k}$.
- The "counting the same thing in two different ways" tool to generate a brand-new formula (which I have never seen in any book so far) for the sum of the first n squares

$$\sum_{i=1}^{n} i^2 = \binom{n+2}{3} + \binom{n+1}{3}.$$

Problem 7.5

A fair coin is tossed repeatedly until there is a run of an odd number of heads followed by a tail. Determine the expected number of tosses.

This was a supplementary problem suggested for the International Mathematical Olympiads, but not included. This is a tough problem, but one that incorporates several of the tools we have developed. Probably the best way for us to start is to just get a coin and do some experimenting. Remember, as soon as we get a run of an odd number of heads, followed by a tail, we have a successful sequence and we quit.

I did the experiment and got the following runs:

Sequence	*Length*
TTHHTTHT	8
HHTHT	5
THHHT	5
TTHT	4
TTHT	4
HT	2
TTHHTHHTHT	10
THT	3
HHHHTHT	7
HHHT	4

The average run length was 5.2.

To actually solve this problem, we will need a couple of insights.

Insight 1

If we denote heads by H and tails by T, any sequence of coin tosses is just a string of Hs and Ts. If we toss a coin n times, we see that each of the tosses can come up as H or T, and there would be a total of 2^n possible arrangements of Hs and Ts in that sequence.

For a sequence to be "successful," it has to have exactly one odd run of Hs followed by a T, which is the last toss. If we let S_n denote the number of successful sequences of length n. then $S_1 = 0$ always, and $S_2 = 1$ is the two-toss sequence consisting of HT.

A successful sequence of length $n \geq 3$ has two possibilities:

1. The first term is H. In this case, the second term must also be H; otherwise, we would have HT, and the sequence would end as a two-toss successful sequence. In this case, the terms HH can then be followed by any other successful sequence of length $n - 2$. This would give us a successful sequence of length n. Thus, the total number of sequences in this category is S_{n-2} (HH, followed by each of the S_{n-2} successful sequences of length $n - 2$).
2. The first term is a T. In this case, the T may be followed by any successful sequence of length $n - 1$. This would again give us a successful sequence of length n. The total number of sequences in this category is S_{n-1} (T, followed by each of the S_{n-1} successful sequences of length $n - 1$).

Thus, we arrive at the equality $S_n = S_{n-1} + S_{n-2}$.

This should look very familiar to us—it is the recurrence relation underlying the Fibonacci sequence. Here, we have $S_1 = 0$ and $S_2 = 1$. Thus, $S_3 = S_2 + S_1 = 1$. Indeed, of the eight possible arrangements in a three-toss sequence (HHH, HHT, HTH, HTT, HHH, THT, TTH, TTT), only THT is a successful sequence. The sequence HTH contains a successful sequence HT, but the tossing would have been terminated there, so it does not count as a successful three-toss sequence.

So, $S_4 = S_3 + S_2 = 2$; $S_5 = S_4 + S_3 = 3$, and so on. If we list some of the S_n terms and the Fibonacci sequence, as we have defined it, we can compare them:

S_1	S_2	S_3	S_4	S_5	S_6	S_7
0	1	1	2	3	5	8
F_1	F_2	F_3	F_4	F_5	F_6	F_7
1	1	2	3	5	8	13

We see that $S_n = F_{n-1}$, where F_n is the nth Fibonacci number. Once again, we see an amazing interrelationship among math problems and how the same ideas keep recurring.

The probability that any sequence of n tosses is a successful sequence is

$$\frac{S_n}{2^n} = \frac{F_{n-1}}{2^n}.$$

As we discussed in Problem 1.7, the expected value in a case like this is calculated as

$$\sum_{i=1}^{\infty} n(p_n),$$

where n is the value of a given term in the summation, and p_n is the probability of getting the value n. Thus, the expected value is the limit L of the sum

$$\sum_{n=2}^{\infty} n(F_{n-1}/2^n).$$

We assume here that this sequence converges (otherwise, the problem has no solution), so we will not delve into a proof of its convergence, and proceed to try to find this sum. There is no easy way to directly sum such a series, so it will need a trick, or more correctly, a good insight, to figure it out.

Insight 2

Assume the sum

$$\sum_{n=2}^{\infty} n(F_{n-1}/2^n)$$

converges to a limit L. In that case, we can say the following. If we start tossing the coin, and the first two tosses are HT, then we have an instant success with a two-toss sequence. The probability of getting HT is 1/4, (1/2 for the first H × 1/2 for the second T). Thus, one-fourth of the time, we need only two tosses, and this product would be 2/4. If we get HH, we expect that we would then need another L tosses to get a successful sequence (we must remember that L is the expected number of tosses to get a success, and the first HH has no bearing on this, because we need some successful sequence to follow the HH). Thus, we would expect, in this scenario, needing $L + 2$ tosses. The probability of starting with HH is 1/4, so one-fourth of time, we need $L + 2$ tosses, and this product is $(L + 2)/4$. Finally, if the first toss is a T, then we expect to need another L tosses to get a success, because a T followed by any successful sequence is itself a successful sequence. The probability of starting with a T is 1/2, so one-half of the time, we need $L + 1$ tosses, and this product is $(L + 1)/2$. We do not need to analyze the possibility of starting with one H because, as we said above, any successful sequence of $n \geq 3$ that starts with one H will need to have a second H and will need to start as HH.

Thus, we can set up an equation for the expected value L, using the terms we have calculated above, since they cover all the possibilities:

$$L = \frac{2}{4} + \frac{L+2}{4} + \frac{L+1}{2}.$$

This is an amazingly clever insight (I can say that because it wasn't my own): we wrote an equation for the expected value in terms of the expected value, and now we can solve for the expected value.

Solving for L, we get $L/6 = 1$ or $L = 6$. Thus, the expected number of coin tosses is 6. This would correspond to the limit of the average if we played this game an infinite number of times, and should be close to the average we would get if we play it a very large number of times.

Tools Used and Developed

- The Fibonacci sequence, which we first encountered in Problem 3.1.
- The definition of expected value, developed in Problem 1.7.
- Writing an equation for the expected value in terms of the expected value.

Problem 7.6

Let p and q be natural numbers such that

$$\frac{p}{q} = 1 - \frac{1}{2} + \frac{1}{3} - \frac{1}{4} + \ldots - \frac{1}{1{,}318} + \frac{1}{1{,}319}.$$

Prove that p is divisible by 1,979.

Note that the number 1,979 figures into this problem because it is from the International Mathematical Olympiad of 1979. (Just because a problem is difficult doesn't mean it can't be cute!) However, since it is no longer 1979, let's instead solve an easier problem: Prove that p is divisible by 17 in

$$\frac{p}{q} = 1 - \frac{1}{2} + \frac{1}{3} - \frac{1}{4} + \ldots - \frac{1}{10} + \frac{1}{11}.$$

This proof can be done directly by getting a common denominator and adding. It is quite tedious, but certainly doable. Doing it this way, the answer comes out to be 20,417/27,720. Checking to see whether 20,417 is divisible by 17, we find that 20,417/17 = 1,201. Thus, we have solved our problem. However, if we want to solve the original 1979 problem, we need an insight.

Insight

We use our insight first to solve the easier version of the problem, and then see that the solution translates directly to the harder problem.

First, we notice that each of the negative terms has an even denominator. This simplifies things for us, because each $-1/(2k)$ can be rewritten as $1/(2k) - 1/k$. Thus, $-1/4 = 1/4 - 1/2$, and so on. This is our first significant insight because it allows us to simplify the series.

Thus, we can rewrite

$$1 - \frac{1}{2} + \frac{1}{3} - \frac{1}{4} + \ldots - \frac{1}{10} + \frac{1}{11}$$

as

$$\left(1 + \frac{1}{2} + \frac{1}{3} + \frac{1}{4} + \frac{1}{5} + \frac{1}{6} + \frac{1}{7} + \frac{1}{8} + \frac{1}{9} + \frac{1}{10} + \frac{1}{11}\right) - 2\left(\frac{1}{2} + \frac{1}{4} + \frac{1}{6} + \frac{1}{8} + \frac{1}{10}\right).$$

This simplifies to

$$\left(1 + \frac{1}{2} + \frac{1}{3} + \frac{1}{4} + \frac{1}{5} + \frac{1}{6} + \frac{1}{7} + \frac{1}{8} + \frac{1}{9} + \frac{1}{10} + \frac{1}{11}\right) - \left(1 + \frac{1}{2} + \frac{1}{3} + \frac{1}{4} + \frac{1}{5}\right),$$

which can be further simplified to

$$\left(\frac{1}{6} + \frac{1}{7} + \frac{1}{8} + \frac{1}{9} + \frac{1}{10} + \frac{1}{11}\right).$$

Our second insight is a clever use of one of the tools we have just developed, the "pairing of terms" tool. We notice that our fractions range from 1/6 to 1/11. We can pair the fractions into pairs of the form

$$\frac{1}{6+j}+\frac{1}{11-j}.$$

Finding a common denominator, we see that each of these "mini-sums" can be written as

$$\frac{(11-j)+(6+j)}{(6+j)(11-j)}.$$

We see that the *j*s cancel out in the numerator, and we are left with a constant sum of 17. Thus, each pair of fractions can now be written as

$$\frac{17}{(6+j)(11-j)},$$

where j will range from 0 to 2.

In other words, what we've done in this simple problem is to pair the fractions as follows:

$$\left(\frac{1}{6}+\frac{1}{11}\right)+\left(\frac{1}{7}+\frac{1}{10}\right)+\left(\frac{1}{8}+\frac{1}{9}\right),$$

to get

$$\frac{17}{6\times 11}+\frac{17}{7\times 10}+\frac{17}{8\times 9}.$$

So now, we have

$$\begin{aligned}\frac{p}{q} &= \frac{17}{6\times 11}+\frac{17}{7\times 10}+\frac{17}{8\times 9}\\ &= 17\left(\frac{1}{6\times 11}+\frac{1}{7\times 10}+\frac{1}{8\times 9}\right).\end{aligned}$$

The fractions in the parentheses can be rewritten as some new fraction p'/q' where q' is the least common denominator of the integers 6 through 11.

The numerator, p', is a complicated sum of the factors left over when we get a common denominator. The nice thing is that we don't even have to bother figuring it out. We can write $p/q = 17p'/q'$, or $17p'q = pq'$. We know that $q' = (6 \times 7 \times 8 \times 9 \times 10 \times 11)$. Each of the factors of q' are relatively prime to 17, which is itself prime. Therefore, 17 cannot divide q'. Thus, it must divide p, and this completes our proof. We saw that when we did this example by brute force, we got $p = 20{,}417$, which indeed is a multiple of 17.

To solve our 1979 problem, we proceed in identical fashion. We write

$$1-\frac{1}{2}+\frac{1}{3}-\frac{1}{4}+\ldots-\frac{1}{1{,}318}+\frac{1}{1{,}319}$$

as

$$\left(1+\frac{1}{2}+\frac{1}{3}+\ldots+\frac{1}{1{,}318}+\frac{1}{1{,}319}\right)-2\left(\frac{1}{2}+\frac{1}{4}+\ldots+\frac{1}{1{,}318}\right),$$

or

$$\left(1+\frac{1}{2}+\frac{1}{3}+\ldots+\frac{1}{1{,}318}+\frac{1}{1{,}319}\right)-\left(1+\frac{1}{2}+\frac{1}{3}+\ldots+\frac{1}{659}\right),$$

which equals

$$\left(\frac{1}{660}+\frac{1}{661}+\ldots+\frac{1}{1{,}318}+\frac{1}{1{,}319}\right).$$

Now we pair our terms as follows:

$$\frac{1}{660+j}+\frac{1}{1{,}319-j},$$

which gives us fractions of the following form:

$$\frac{(1{,}319-j)+(660+j)}{(660+j)(1{,}319-j)},$$

or

$$\frac{1,979}{(660+j)(1,319-j)}.$$

Thus, we have paired our fractions in the following way

$$\left(\frac{1}{660}+\frac{1}{1,319}\right)+\left(\frac{1}{661}+\frac{1}{1,318}\right)+\ldots+\left(\frac{1}{989}+\frac{1}{990}\right),$$

to get

$$\frac{p}{q}=\frac{1,979}{660\times 1319}+\frac{1,979}{661\times 1318}+\ldots+\frac{1,979}{989\times 990}=1,979\frac{p'}{q'},$$

where q' is the product of all the integers from 660 to 1,319. Each of these is relatively prime to 1,979, which is itself prime. Thus, $1,979p'q = pq'$, and since 1,979 does not divide q', it must divide p.

We see that the key to solving this problem was a clever use of our pairing tool in summing a series.

Tools Used and Developed

- The pairing tool when summing a series, developed in Problem 1.3.
- In a series of the form $1 - 1/2 + 1/3 - 1/4$, , each $-1/(2k)$ can be rewritten as $1/(2k) - 1/k$.

Problem 7.7

The first n positive integers $(1, 2, 3, \ldots, n)$ are spotted around a circle in any order, and the positive differences $d1, d2, \ldots, dn$ between consecutive pairs are determined. Prove that, no matter how the integers may be jumbled around the circle, the sum of these n differences,

$$S = d1 + d2 + \ldots + dn,$$

will always amount to at least $2n - 2$.

If the integers around the circle are $a1$, $a2$, … , an, the problem involves the sum of the absolute values

$$|a1 - a2| + |a2 - a3| + |a3 - a4| + \ldots. + |an - a1|$$

Here is an example for $n = 9$.

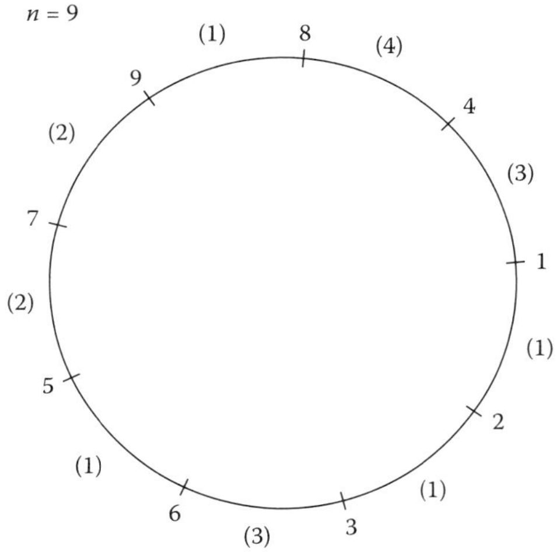

Working with absolute values is somewhat difficult. In Professor Elkies's solution, he saw that both n and 1 occur somewhere on the circle, so these can be used to divide the circle into two arcs. Let's look at one of these arcs.

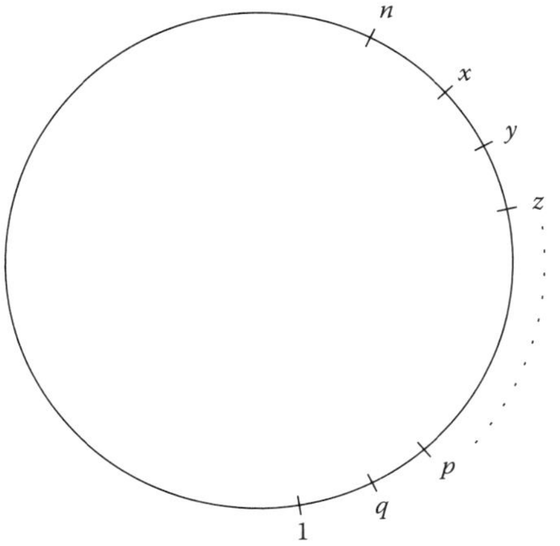

Now we'll sum the differences that lie along that arc:

$$S1 = |n - x| + |x - y| + |y - z| + \ldots + |p - q| + |q - 1|.$$

We now use the well-known (and logical) identity about absolute values that

$$|a + b + c + \ldots| \leq |a| + |b| + |c| + \ldots.$$

Thus, letting $a = n - x$, $b = x - y$, $c = y - z$, etc., we have

$$\begin{aligned} S1 &= |n - x| + |x - y| + |y - z| + \ldots + |p - q| + |q - 1| \\ &\geq |n - x + x - y + y - z + \ldots + p - q + q - 1|. \end{aligned}$$

When we apply the magic of telescoping to the right-hand side, we get $S1 \geq n - 1$.

There was nothing special about the arc we picked. We can apply the exact same argument to the other arc containing n and 1 and call that sum $S2$. Thus, automatically, we have $S2 \geq n - 1$.

Since $S = S1 + S2$, we have the desired result, $S \geq 2n - 2$.

Simple enough. But let's consider an insight I had that helped me solve this problem in a different way.

Insight

Let's approach this problem by using induction. First, note that for the case $n = 2$, we get a sum of $S = 2$, which satisfies the requirement that $S \geq 2n - 2$. Next, we show that if this holds for the case of n numbers around a circle it must also hold for the case of $n + 1$ numbers positioned around the circle.

We consider the numbers 1 through n placed around the circle. To create the case in which the numbers 1 through $n + 1$ are around the circle, we simply place the number $n + 1$ between any two of the numbers on our circle (call these numbers a and b, and let $a > b$).

In placing $n + 1$ between a and b, we have added two new terms, namely $[(n + 1) - a]$ and $[(n + 1) - b]$, to our total sum, and we have eliminated one term, $(a - b)$, from our sum. By combining these two effects, we thus have

$$S(n + 1) = S(n) + (n + 1 - a) + (n + 1 - b) - (a - b),$$

which simplifies to

$$S(n+1) = S(n) + 2(n+1-a).$$

To find the minimum value of $S(n+1)$, we can plug in the largest value possible for a. Because a was a number on our original circle, the largest possible value for a is $a = n$. This gives us $S(n+1) = S(n) + 2$. Therefore, $S(n+1) \geq S(n) + 2$, and if $S(n) \geq 2n - 2$, we get

$$S(n+1) \geq S(n) + 2 \geq 2n = 2(n+1) - 2.$$

Tools Used and Developed

- Mathematical induction, developed in Problem 1.3.

Problem 7.8

Imagine that 12 students stand in a circle, and one person is designated as "number 1." We go around the circle, and eliminate every second student, and keep cycling this way until one student remains, who is designated as the winner. If a student wants to win, where should he or she stand?

This exercise is a variation of a famous problem in mathematics known as the Josephus problem, named for the first century Jewish historian, Josephus Flavius. As the legend goes, Josephus and a group of 40 other Jewish soldiers were besieged by the Roman army. Preferring death to surrender, they agreed to stand in a circle, and go around with every third man committing suicide until they all died. Josephus and a compatriot were not on board with this plan, and so Josephus, using his mathematical skills, quickly calculated where he and his friend should stand so that they were the last two left alive. Thus, Josephus survived, surrendered to the Romans, and went on to become a famous historian.

Now, to solve this problem, let's just draw the students in a circle, number them 1 through 12, and then go through the sequence of events as outlined in the problem.

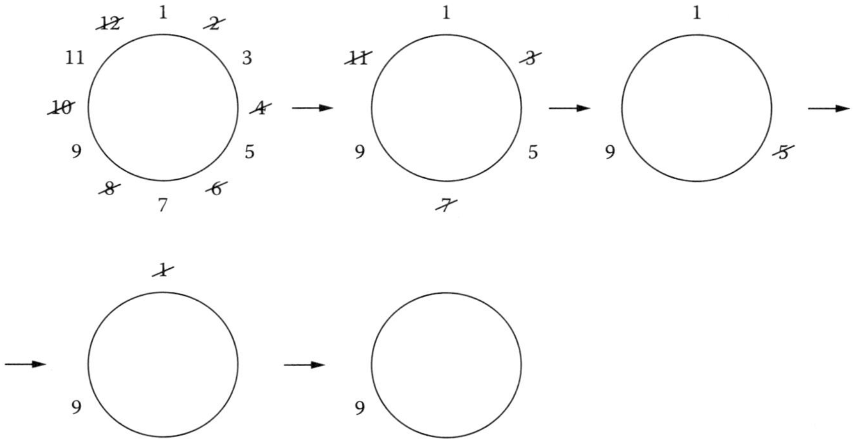

In the first pass, we eliminate every second student, meaning that we eliminate students 2, 4, 6, 8, 10 and 12. So students 1, 3, 5, 7, 9, and 11 are still in the circle. The last person eliminated was 12, so the next person to be eliminated is 3 (we skip student 1 because we continue around the circle eliminating every second person, and student 3 is now the second person after we have eliminated 12). Then, we skip 5 and eliminate 7, skip 9 and eliminate 11. Now only 1, 5 and 9 are left. We eliminated 11, so we skip 1 and eliminate 5, leaving 9 and 1. Since 5 was the last person eliminated, we skip 9 and eliminate 1, leaving 9 as the last person standing—the winner. Thus, we can say $J(12) = 9$. In other words, for our modified Josephus problem, as we have stated above, if we have 12 people, the last one standing is person 9.

Now let's do this problem for 48 people, and then for 100 people. In other words, let's figure out $J(48)$ and $J(100)$. This could get very tedious. We need some sort of insight.

Insight 1

We see that $J(k)$ will be odd because we eliminate all the even-numbered people in the first pass. If we start with an even number of people, say $2n$, then after the first pass, we have the result shown in the following diagram.

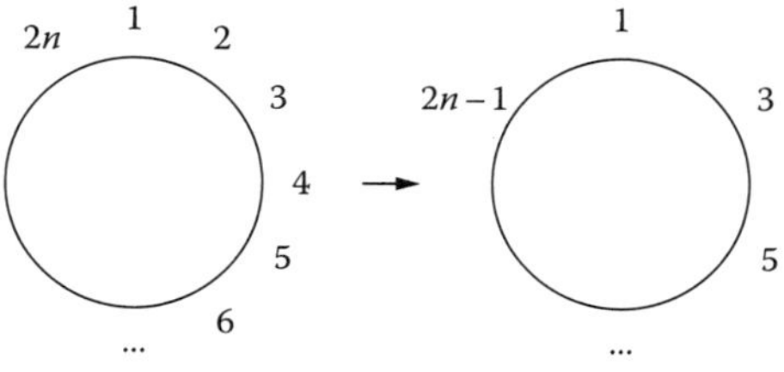

Now we have n people left, and number 3 will be the first to be eliminated. Therefore, this is just like the original game starting with n people, except that each person's number has been doubled and then decreased by 1. For example, if when we start with n people, the third person is the winner, we see from the above circle on the right that this is person 5 from the original circle with $2n$ people; that is, $2(3) - 1 = 5$. We can generalize this as follows. If $J(n) = k$, then that person would have been number $2k - 1$ in the original circle of $2n$ people. Thus, we can say that

$$J(2n) = 2J(n) - 1.$$

This gives us a recursive formula for $J(2n)$ in terms of $J(n)$ for an even number of people starting the game. Therefore, we can immediately answer our first question, to find $J(48)$. From the above, we know that $J(24) = 2J(12) - 1$. Since we found that $J(12) = 9$, we have $J(24) = 2(9) - 1 = 17$. Similarly, $J(48) = 2J(24) - 1 = 2(17) - 1 = 33$. Thus, the last person standing is number 33.

Insight 2

Clearly, just doubling values starting from 12 will not get us to $J(100)$. Moreover, it would be great to have a general solution. Therefore, let us try to deal with all cases. Previously, we got a recursive formula for an even number ($2n$) of starting players. Now, let's deal with an odd number of players, $2n + 1$.

In this case, in the first round we eliminate all of the even players: 2, 4, 6, … $2n$. This means n players are eliminated so far, leaving $n + 1$ players in the circle. The next person to be eliminated will be player 1 (right after $2n$, we skip $2n +1$ and eliminate 1). This now leaves n odd-numbered players in the circle, starting with person 3, as shown below.

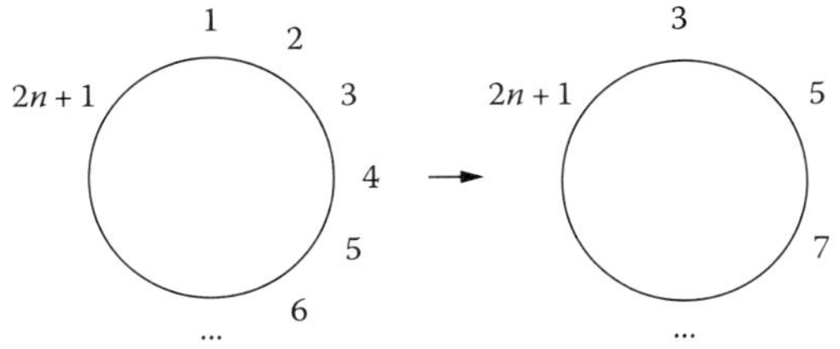

This is very similar to the previous case, where the situation had been reduced to an n-person game. Now, though, we see that if in this n-person game, the kth person is the winner, this would correspond to person $2k + 1$ in the original $2n + 1$ circle. Thus, we have established that for an odd number of players $2n + 1$, that $J(2n + 1) = 2J(n) + 1$.

Lastly, we see that if we have a 1-person game, that person survives, so $J(1) = 1$. Putting all of these together, we now have a general recursion for J that covers all cases:

$$J(1) = 1,$$

$$J(2n) = 2J(n) - 1,$$

$$J(2n+1) = 2J(n) + 1.$$

This is a bit different from a typical recursion in which $J(n)$ is some function of $J(n - 1)$, but it still lets us quickly build up a table for small values. For example, $J(2) = 2J(1) - 1 = 1$. This makes sense: if we have two people, and eliminate every second person, then person 2 gets eliminated right away, leaving person 1. For three people, we would eliminate person 2, skip person 3, and eliminate person 1, leaving person 3. This jibes with the above formula: For three people, $n = 1$ in the formula $J(2n + 1) = 2J(n) + 1$; thus, we have $J(3) = 2J(1) + 1 = 3$. Proceeding in this way, we get the following table.

n	1	2	3	4	5	6	7	8	9	10	11	12	13	14	15	16
$J(n)$	1	1	3	1	3	5	7	1	3	5	7	9	11	13	15	1

We can see the solution pattern now. We know $J(n)$ is always odd (because the even people get eliminated on the first pass, as we said before). Also, for every $n = 2^k$, we get $J(n) = 1$, and then we move up the consecutive odd numbers until $n = 2^{k+1}$, and start at 1 again. Therefore, if we write $n = 2^m + r$, where 2^m is the largest power of 2 not exceeding n, and r is the remainder, then

$$J(2^m + r) = 2r + 1, \quad \text{with} \quad 0 \le r < 2^m.$$

Of course, this is just a guess, or a conjecture, based on the pattern we are seeing and understanding why the pattern would continue. For example, since $J(4) = 1$, and when the input to J is even, $J(2n) = 2J(n) - 1$, then we know that $J(8) = 1$. Then we also know that $J(16) = 1$, and so on for all powers of 2. We can also calculate $J(9)$ from $J(4) = 1$ and $J(2n + 1) = 2J(n) + 1$ for odd input, giving us $J(9) = 3$. Similarly, $J(17) = 2J(8) + 1$, giving us $J(17) = 3$.

Working in this way, we have very good reason to believe that our conjecture is true. It would be nice, however, to have a formal proof rather than just this conceptual, albeit convincing, argument. We can prove our conjecture by induction.

We want to prove that if $n = 2^m + r$ and $0 \le r < 2^m$, then $J(2^m + r) = 2r + 1$. We know the case is true for $n = 1$. In that case, $m = 0$ and $r = 0$. So $J(1) = 1$. Now, we deal with the cases of even input to J and then odd input to J.

For even n, we assume that we know the theorem is true for $n/2$, where we have

$$\frac{n}{2} = 2^{m-1} + \frac{r}{2}.$$

For even input, we know that $J(2n) = 2J(n) - 1$. We put $n/2$ in place of n, and we have, $J(n) = 2J(n/2) - 1$. However, from the induction hypothesis, $J(n/2) = 2(r/2) + 1$, so

$$J(n) = 2\left[2\left(\frac{r}{2}\right)+1\right]-1.$$

Thus, $J(n) = 2r + 1$.

For odd n, since $n = 2^m + r$, we know r is also odd. Let's choose r_1 such that $(n - 1)/2 = 2^{m-1} + r_1$. We note that $r_1 = (r - 1)/2$. For odd n, we know that $J(n) = 2J[(n - 1)/2] + 1$. From the induction hypothesis, $J[(n - 1)/2] = 2r_1 + 1$, so $J(n) = 2(2r_1 + 1) + 1$. By substituting $r_1 = (r - 1)/2$, we have $J(n) = 2[(r - 1) + 1] + 1 = 2r + 1$. This completes the proof.

Now we are ready to tackle the unsolved portion of our problem, which is $J(100)$. We know that 100 can be written as $100 = 2^6 + 36$. Therefore, $r = 36$, and since $J(n) = 2r + 1$, we have $J(100) = 2(36) + 1 = 73$. Therefore, to avoid elimination (or death, as in the case of Josephus), one should stand in position 73.

This problem could have kept us busy for a long time, with a potential for miscalculation along the way—which could have had dire consequences.

Tools Used and Developed

- Using an insight based on case work for small values and recognizing a pattern.
- Proof by mathematical induction, such as $J(n) = 2r + 1$.
- Setting up a recursion relation and learning how to break down even and odd cases, such as $J(2n) = 2J(n) - 1$ for even inputs, and $J(2n + 1) = 2J(n) + 1$ for odd inputs.

Problem 7.9

Prove the following statement:

$$n^n > 1 \times 3 \times 5 \times \cdots \times (2n-1) \qquad \text{for } n > 1.$$

In other words, for any positive integer n greater than 1, if we raise n to the nth power, the result will be greater than the product of the first n odd integers. For example, if $n = 2$, then 2 raised to the second power (4) is greater than the product of the first two odd integers (1×3). We want to prove that this holds across the board. But how?

As a general rule, whenever we have to prove a statement for all integers, one good strategy to at least consider is mathematical induction (which we introduced in Problem 1.3). Here, we already have a base case for $n = 2$. Then, by induction, we want to prove that if it is true for n, it will be true for $n + 1$.

Thus, we want to prove that if $n^n > 1 \times 3 \times 5 \times \ldots \times (2n - 1)$, then $(n + 1)^{n+1} > 1 \times 3 \times 5 \times \ldots \times (2n - 1)(2n + 1)$, where the left-hand side now represents the product of the first $n + 1$ odd numbers.

This may be a bit of a tough proof if we haven't done anything like it before, since we really have a difficult time knowing where to start. If we just start thinking about $(n + 1)^{n+1}$, however, we can see that if we were to expand this binomial by using the binomial theorem (which we've already learned about in Problem 2.6), then we would get the following terms:

$$n^{n+1} + \binom{n+1}{1} n^n 1^1 + \binom{n+1}{2} n^{n-1} 1^2 + \cdots.$$

One of the hardest things in math is knowing what we can ignore. Here, since we know something about n^n, we can try to work with terms that include it, ignoring terms that come after that with lesser powers of n. Remembering that 1 raised to any power is 1, and that $\binom{n+1}{1} = n + 1$, we can then say that $(n + 1)^{n+1} = n^{n+1} + (n + 1)n^n$ + some other stuff (i.e., the rest of the binomial expansion). If we can prove that $n^{n+1} + (n + 1)n^n$ is greater than $1 \times 3 \times 5 \times \cdots \times (2n - 1)(2n + 1)$, then we will have succeeded, because $(n + 1)^{n+1}$ is even bigger than $n^{n+1} + (n + 1)n^n$.

Now, let's use a little trick. Let us rewrite n^{n+1} as $n \times n^n$. We have, $n^{n+1} + (n + 1)n^n = n \times n^n + (n + 1)n^n$. Factoring out n^n, we have $n^{n+1} + (n + 1)n^n = n \times n^n + (n + 1)n^n = n^n(2n + 1)$ Since $n^n > 1 \times 3 \times 5 \times \ldots \times (2n - 1)$, then $n^n(2n + 1) > 1 \times 3 \times 5 \times \ldots \times (2n - 1)(2n + 1)$, and, since $(n + 1)^{n+1} > n^{n+1} + (n + 1)n^n = n^n(2n + 1)$, then $(n + 1)^{n+1} > 1 \times 3 \times 5 \times \ldots \times (2n - 1)(2n + 1)$, completing our proof.

This proof was a bit tricky because we had to deal with the binomial expansion and then had to know or figure out that we could ignore a lot of the terms and work with only the first few.

Insight

A shorter and more elegant proof uses some material that we have already learned that has to do with n^n and the product of the first n odd numbers. Let's start with the relationship of interest: $n^n > 1 \times 3 \times 5 \times \ldots \times (2n - 1)$. We get rid of the exponential by taking the nth root of both sides, so now we have

$$n > [1 \times 3 \times 5 \times \cdots \times (2n-1)]^{1/n}. \qquad (7.1)$$

Now comes the insight. On the left-hand side of Equation (7.1), let's do something that seems odd. Let's rewrite n as n^2/n. Initially, this seems like it just complicates things, but if we recall the relationship we developed in Problem 1.3, where we wanted to find a compact way to calculate the sum of the first n odd numbers, we replaced $1 + 3 + 5 + \ldots + (2n - 1)$ with n^2. Now we're going to go the other way and rewrite n^2/n as $[1 + 3 + 5 + \ldots + (2n - 1)]/n$, which, of course, just equals n.

Therefore, Equation (7.1) can now be rewritten as

$$\frac{1+3+5+\ldots+(2n-1)}{n} > [1 \times 3 \times 5 \times \cdots \times (2n-1)]^{1/n}. \qquad (7.2)$$

This is the relationship we want to prove. But look—it's already done! The left-hand side of Equation (7.2) is the arithmetic mean of the first n odd numbers, and the right-hand side is the geometric mean of the first n odd numbers. By the AM-GM inequality (which we covered in Problem 6.9), we know that the arithmetic mean will be greater than the geometric mean, unless all of the summands are equal (which in this case, since we are adding different odd numbers, they are not). Therefore, the proof just falls right out for us.

Tools Used and Developed

- $1 + 3 + 5 + \ldots + (2n - 1) = n^2$ (developed in Problem 1.3).
- AM-GM inequality (developed in Problem 6.9).
- We can't have an insight to use a particular tool if we don't even know that the tool exists.

Epilogue and Acknowledgments

My pretense for this book is that each of us is the young Gauss in Master Buttner's classroom and that we are presented weekly with math problems that seem to exist just to keep us busy. However, at the end of our journey, after we have solved all of the "busy work" problems, we realize something remarkable: tackling these problems was actually the funnest part of math class. Moreover, I think we learned an amazing amount of math in the process and gained a new appreciation for the process of mathematical insight. I have come to realize that Master Buttner was not lazy or mean at all—he actually was a very wise and skilled educator who has taken me to mathematical places I would have never gone on my own. I owe him a great debt and have begun to think that perhaps it is time to start rehabilitating the image of poor Master Buttner!

The above was in the realm of fiction and pretense. In reality, however, I owe acknowledgements to many and would like to mention a few of them here. First and foremost, I would like to acknowledge my older brother Leith, who first showed me the magic of mathematical insight. I would also like to thank my outstanding math teachers at Flintridge Preparatory School; their passion for math was infectious. Also, although I have benefitted tremendously from all of the books I listed in the sources section to follow, I am especially grateful for the wonderful *The Art of Problem Solving* books written by Richard Rusczyk; I have never encountered their equal, and they were the source not only of inspiration but also of many problems. Similarly, *The Art and Craft of Problem Solving* by Paul Zeitz is another outstanding source. Both of these really teach how to think! I would also like to acknowledge Dr. Klaus Peters of A K Peters Publishing for giving me the opportunity to write this book and for believing that a young person could make a contribution. Finally, I would like to thank the absolutely wonderful team at Taylor and Francis, including Sunil Nair, Charlotte Byrnes, and Sandra Rush, for their support and all-around excellence.

Problem and Figure Credits

Problem 1.4, including the figures, is reproduced with permission from the 2000 American Mathematics Competition (AMC) 12 contest (problem 8). The solution and figures appear in J. Douglas Faires, *First Steps for Math Olympians: Using the American Mathematics Competitions*, MAA Problem Book Series, Mathematical Association of America, 2006. Copyright the Mathematical Association of America, 2012. All rights reserved.

Problem 3.2 was the last problem (problem 25) on the 2010 American Mathematics Competition (AMC) 8 contest.

Problem 5.3: The original version of this problem appeared in the American High School Mathematics Examination (AHSME). The problem has been modified a bit here to make it more friendly.

Problem 5.4 is from the 2008 American Invitational Mathematics Examination (AIME) II exam (problem 8). The problem has been modified a bit here to make it a little easier.

Problem 6.10 comes from the excellent book *More Mathematical Challenges* by Tony Gardiner (1997). It was used in the 1995 UK Junior Mathematical Olympiad. The figures in this problem are from pages 8, 13, 69, and 77 of that publication, reprinted with permission of Cambridge University Press.

Problem 7.1 is adapted from the 1988 Spanish Olympiad.

Problem 7.2 is adapted from the 1991 Leningrad Mathematical Olympiad, problem 18.

Problem 7.3: Figures are reproduced with permission from Leith Hathout, "Iterating the Number of Intersection Points of the Diagonals of Irregular Convex Polygons, or C(n,4) the Hard Way!" *Teaching Mathematics and Its Applications* 26:1 (March 2007), 38–44,doi: 10.1093. © Leith Hathout, 2006.

Problem 7.6 is from the International Mathematical Olympiad of 1979.

Problem 7.7 is from the mathematical journal *Kvant*, where the solution was provided by the world-famous problem-solver Naom Elkies while he was a student at Columbia University. The figures are from Honsberger (1990), p. 148. © Mathematical Association of America, 2012. All rights reserved.

Sources and Suggested Reading

Andreescu, Titu, and Bogdan Enescu. *Mathematical Olympiad Treasures*. Birkhauser, Boston, 2004.

Andreescu, Titu, and Razvan Gelca. *Mathematical Olympiad Challenges*. Birkhauser, Boston, 2000.

Coxford, Arthur and Joseph Payne. *Advanced Mathematics: A Preparation for Calculus, 2nd edition*. Harcourt Brace Jovanovich, Inc., New York, 1978.

Faires, J. Douglas. *First Steps for Math Olympians*. Mathematical Association of America, Washington, DC, 2006.

Gardiner, Tony. *More Mathematical Challenges*. Cambridge University Press, Cambridge, UK, 1997.

Honsberger, Ross. *In Polya's Footsteps*. Mathematical Association of America, Washington, DC, 1997.

Honsberger, Ross. *More Mathematical Morsels*. Mathematical Association of America, Washington, DC, 1990.

Klamkin, Murray S. *International Mathematical Olympiads 1978-1985*. Mathematical Association of America, Washington, DC, 1999.

LeithHathout, "Iterating the Number of Intersection Points of the Diagonals of Irregular Convex Polygons, or C(n,4) the Hard Way!" *Teaching Mathematics and Its Applications* 26(1) (March 2007), 38–44.

Moser, Leo, Problems and Questions, Q. 87, *Mathematics Magazine* 26 (March 1953), 226.

Nahin, Paul. *Digital Dice: Computational Solutions to Practical Probability Problems*. Princeton University Press, Princeton, NJ, 2008.

Nahin, Paul. *Duelling Idiots and Other Probability Puzzlers*. Princeton University Press, Princeton, NJ, 2000.

Niven, Ivan. *Mathematics of Choice: How to Count without Counting*. Mathematical Association of America, Washington, DC, 1965.

Poonen, Bjorn, and Michael Rubinstein, *Th Number of Intersection Points Made by the Diagonals of a Regular Polygon. SIAM J. Discrete Math. 11* (1998), 135–156.

Posamentier, Alfred S., and Charles T. Salkind. *Challenging Problems in Algebra*. Dover Publications, New York, 1996.

Rusczyk, Richard. *The Art of Problem Solving: Intermediate Algebra*. AoPS Incorporated, Alpine, CA, 2009.

Rusczyk, Richard. *The Art of Problem Solving: Precalculus*. AoPS Incorporated, Alpine, CA, 2009.

Savchev, Svetoslav, and Titu Andreescu, *Mathematical Miniatures*. Mathematical Association of America, Washington, DC, 2003.

Trigg, Charles. *Mathematical Quickies*. Dover Publications, New York, 1985.

Vakil, Ravi. *A Mathematical Mosaic: Patterns and Problem Solving*. Brendan Kelly Publishing, Inc., Burlington, Ontario, 1996.

Zeitz, Paul. *Th Art and Craft of Problem Solving*. John Wiley & Sons, Inc., New York, 1999.

Index